Ali AROUS
Faiza Nawel GHOMARI

Effects of water stress on the Algerian cereal industry

Ali AROUS
Faiza Nawel GHOMARI

Effects of water stress on the Algerian cereal industry

ScienciaScripts

Imprint

Any brand names and product names mentioned in this book are subject to trademark, brand or patent protection and are trademarks or registered trademarks of their respective holders. The use of brand names, product names, common names, trade names, product descriptions etc. even without a particular marking in this work is in no way to be construed to mean that such names may be regarded as unrestricted in respect of trademark and brand protection legislation and could thus be used by anyone.

Cover image: www.ingimage.com

This book is a translation from the original published under ISBN 978-620-6-72732-3.

Publisher:
Sciencia Scripts
is a trademark of
Dodo Books Indian Ocean Ltd. and OmniScriptum S.R.L publishing group

120 High Road, East Finchley, London, N2 9ED, United Kingdom
Str. Armeneasca 28/1, office 1, Chisinau MD-2012, Republic of Moldova, Europe
Printed at: see last page
ISBN: 978-620-8-30995-4

EUROPEAN UNIVERSITY PUBLISHING.
EFFECTS OF WATER STRESS ON THE CEREALS SECTOR IN ALGERIA - THE CASE OF DURUM WHEAT IN SEMI-ARID ZONES
DR. AROUS ALI

SUMMARY

The phases of grain formation for the acquisition of its weight and quality are considered to be decisive stages in the development of yield in durum wheat (Triticum durum Desf.). The aim of the present work is to study grain filling in wheat under different water regimes. To this end, trials involving 120 genotypes were carried out in the field over three consecutive growing seasons and involved evaluating the remobilisation of reserves and the current photosynthesis of different organs on this process. The results show that grain filling is greatly influenced by genetic variability and the water regime adopted. Under rainfed conditions, where plants are subject to variable water deficits, the remobilisation of reserves and photosynthetic activity of the components of the ear constitute the main assimilates essential for the accumulation of reserves in the grain. The neck of the ear, with its long life and position close to the ear, is also important for the availability of the photo-assimilates needed for grain formation, thanks to its photosynthetic activity. Under irrigated conditions, where the optimum water supply increases the longevity of the organs located above the last node, it accentuates their contribution to the final weight of the grain, particularly that of the leaf system. However, water-deficient situations significantly reduce the morphological traits of the stem part of the plant. In order to distinguish the effect of variations in water supply alone on the grain filling process and the development of its qualities, two trials were carried out under controlled conditions. They involved five genotypes grown under three water regimes, 100%CC, 60%CC and 30%CC. The results show that variations in water regimes lead to very significant variations in the relative contributions of the various plant organs to grain filling. They show that the accentuation of the water deficit is accompanied by a clear contribution of the remobilisation of reserves and photosynthesis of the ear to grain filling. Grain weight evolution indices and growth kinetics depended on the water regime and organs. In the local and Waha genotypes, the remobilisation of reserves maintains an increasing kinetic with the accentuation of the water deficit, whereas in the introduced genotypes, current photosynthesis appears to be more efficient under these conditions.

Key words: durum wheat; grain filling; remobilisation; current photosynthesis; organs; water deficit.

TABLE OF CONTENTS

INTRODUCTION

Drought is one of the main factors affecting the productivity of cultivated species. Indeed, the constraints generated by this abiotic stress greatly limit the production of various crops in many countries around the world (Kamoshita et al., 2008; Lonbani, 2011). According to Witcombe et al. (2008), thirty-five percent (35%) of the world's arable land is classified as arid or semi-arid, where the risk of water shortage will become increasingly frequent and persistent in the future as a result of climate change. In Mediterranean countries, particularly those on the southern shores, water shortages have always been a serious obstacle to improving yields from rain-fed crops (Kamoshita et al., 2008).

In Algeria, the location of cereal growing in the country's high plateaux and inland plains exposes these species to unfavourable climatic conditions that are detrimental to the expression of their productive potential (Adda et al., 2013).

Among cereal species, durum wheat (Triticum durum Desf.) is of primary economic interest in crop production in Algeria. The products of this crop are major elements in the dominant consumption pattern of the local population. Nevertheless, local production of this product remains largely insufficient to meet expressed needs, which continue to grow year on year. This is an obligation that makes our country one of those dependent on world markets. It should be noted in this context that our country is currently one of the world's leading importers of durum wheat, capturing almost 50% of the world market; it imported 545,000 tonnes of durum wheat at the end of November 2012 (CIC, 2013). The main reason for this situation is the low yields recorded in the various marketing years, which are the result of a number of constraints, particularly climatic. Indeed, like all rainfed crops in Algeria, the low yields recorded for durum wheat reflect the constraints caused by the water deficit. This abiotic stress is the main factor responsible for variations in the quantity and quality of the durum wheat crop. Variations in the intensity and perseverance of drought during the development cycle of this cereal produce versatile impacts on its productivity.The sensitivity of the yield development process in durum wheat to water deficit is particularly high during the grain filling period (Saeedipour and Moradi, 2011). This phenomenon is aggravated by the fact that this stage at the end of the development cycle takes place at high temperatures (Chenaffi et al., 2006). Grain weight and quality are the reference characteristics for assessing yield (Gate, 1995) and therefore productivity in durum wheat. The grain filling process depends on the supply of carbon products from two main sources, the translocation of reserves stored in the plant and the routine production of photoassimilates by its various organs (Blum et al., 1991; Yang and Zhang 2006; Zhu et al., 2010). It should be noted that the deployment of these resources through the unfolding of processes ensuring their availability is strongly influenced by water availability during the heading phase. The declaration of a water deficit during this period depends on its severity, which proportionally reduces the quantity of grain produced, as well as its quality. The effects of inadequate water supply during this period are felt by the vegetative and reproductive organs involved in grain formation and the production of the photoassimilates essential for grain filling. The grain filling process is carried out jointly by current photosynthesis and the translocation of reserves from the various stem organs (Cruz et al., 2000). Current photosynthesis emanates mainly from the organs located above the last node. These include the last leaf, the neck of the ear, the last internode, the beard and the developing grain and its husks. The involvement of these organs depends on their longevity, which is conditioned by the quality of the water supply (Araus et al., 1993; Maydup et al., 2010). The translocation of assimilates, mainly fructose during the post-anthesis period, takes place from reserves accumulated in the stem (Foulkes et al., 2007; Ehdaie et al., 2008; Álvaro et al., 2008). The contribution made by current photosynthesis by certain organs is very marked, as in the case of the last leaf, which is considered to be essential in determining grain yield due to its active duration during grain filling (Abbad et al., 2004; Khaliq et al., 2008). The involvement of other organs such as the barbs, the neck of the ear and the husks of the grain in the availability of assimilates has been less studied. Water deficit significantly alters the duration of photosynthetic activity in these organs, the formation of reserves and the acuity of the sap circulation responsible for their migration into the developing grains.The constraints are expressed by a reduction in photosynthetic activity, inhibiting the various metabolic reactions (Araus et al., 2002; Li et al., 2003; Tambussi et al., 2005), and by a reduction in the duration of activity of the various organs responsible, limiting their longevity. As far as the translocation of reserves is concerned, the depletion of water in the plant is detrimental to the circulation of sap, which is essential for the movement of assimilates. Under these conditions, the effective contribution of the various processes involved in filling the grain depends on its behaviour with regard to the water deficit during this period. These parameters include, essentially, the proximity of the organ to the grain and its photosynthetic yield, the longevity of the organs and the hydraulic resistance of the conductive tissues to ensure the circulation of the sap. Estimating the contribution of the various processes to grain formation in durum wheat is poorly understood and difficult to measure, particularly in the field. Grain formation and the development of its weight and quality require a thorough understanding of the processes involved. An in-depth study of the contribution of the various organs in these processes and their sensitivity to water deficit is essential for the selection of accessible criteria in programmes to create drought-tolerant cultivars. The objectives of the research carried out in the present work are in line with this strategy. They make it possible to evaluate the contribution of the current photosynthesis shares of the various plant organs and the remobilisation of reserves stored during vegetative activity, in the grain filling process and the building up of its quality. A diversified genetic variability comprising several genotypes with divergent origins, phenology, phenotypic expression and drought sensitivity indices was used to carry out the present work. The study involved trials conducted under rainfed conditions over three consecutive seasons and those conducted under controlled conditions where the effects of different water regimes on grain filling were assessed. The distinction between the contributions of the various plant organs through their photosynthetic activity and the remobilisation of reserves was studied.

BIBLIOGRAPHIC SYNTHESIS

1. Presentation of the species studied

1.1. General information on wheat durum

Historically, wheat was one of the first cereals grown in the world. In quantitative terms, it is the third most widely grown cereal, with around 600 million tonnes grown each year (Clerget, 2011). It was domesticated in the Near East from a wild grass around 10,000 years ago (Naville, 2005). Its consumption dates back to ancient times. The first crops appeared in Mesopotamia and in the valleys of the Tigris and Euphrates rivers (now Iraq), in the region of the "fertile crescent" (present-day Lebanon, Syria and southern Turkey), where wild wheat has survived to this day. All spontaneous and cultivated wheat species belong to the genus Triticum and are distributed over a vast area stretching from Central Asia to the Mediterranean basin (Verville, 2003). Wheat reached Western Europe via two major routes, the Mediterranean and the Danube valley (Naville, 2005).

1.2. Features morphological

Wheats are medium-height grasses that can reach up to 1.5 m depending on the variety. They have alternate, flattened leaves, consisting of a base (sheath) surrounding the stem, a terminal part that aligns with the parallel veins and a pointed tip. At the point of attachment of the leaf sheath is a thin, transparent membrane (ligule) with two small lateral appendages (auricles). The root system comprises seminal roots produced by the seedling during emergence, and adventitious roots which form later from the nodes at the base of the plant and constitute the permanent adventitious root system. Durum wheat has an erect, hollow or solid cylindrical stem (stubble) subdivided into internodes (Clarke et al., 2002). The spikelets at the end of the culms are made up of small, inconspicuous flowers. The flowers have no petals and are surrounded by two glumes. Each cleistogamous flower contains three stamens and an ovary topped by two feathery styles. After fertilisation, the ovary gives rise to the seed, which is both the fruit and the seed or caryopsis (Mosiniak et al, 2006; Boutigny, 2007).

1.3. Features botanicals

Wheat is an annual herbaceous monocotyledonous plant belonging to the genus Triticum of the grass family (Poaceae). Today, there are several species of wheat, two of which dominate production: common wheat (Triticum aestivum) and wheat (Triticum aestivum). durum wheat (Triticum durum), distinguished by their number of chromosomes. Durum wheat has a tetraploid genome (AABB). Each genome is made up of 7 pairs of homologous chromosomes, giving a total of 28 chromosomes ($2n = 4x = 28$).

According to Feillet (2000), Le Clech (2000) and Naville (2005), the taxonomy of durum wheat is as follows:
Class: Angiosperms

Order: Monocotyledons

Family: Poaceae (Gramineae)

Tribe: Hordeae
Genus: Triticum

Species: Triticum durum

Common name: Durum wheat

1.4. Economic importance of wheat

Wheat is one of the three major "cereals" along with maize and rice (Jacquemin, 2012). Worldwide production reached 650 million tonnes in 2010 (Branlard et al., 2012). According to studies and projections published by the FAO, which take into account global demographic development, requirements are set to increase further in the coming years, probably reaching 1,000 million tonnes of wheat in 2020. Increased production will inevitably be achieved by improving productivity, given that the area set aside for growing wheat, estimated at 220 million hectares, cannot grow indefinitely, especially as agro-climatic and growing conditions (semi-arid zones, soil salinity) and even the extension of urban areas are real limits to its progress (Branlard et al., 2012). Wheat is mankind's most important food resource, and the main source of protein. It is also a key resource for animal feed and a wide range of industrial applications (Bonjean and Picard, 1991; Doré and Varoquaux, 2006).Algeria is one of the world's leading wheat-consuming countries. As a result of the imbalance between production and consumption mentioned above, our country is one of the world's top three importers of cereals, after Brazil and Egypt (Kellou, 2008). Durum wheat grain has a high nutritional value, due to its high protein content and the presence of gluten, which makes pasta more resistant to cooking (Tab.1). Cereal derivatives are also used in many non-food products such as medicines, paper, textiles, glues, detergents, paints, plastics and now biofuels ("green fuels") (Boutigny, 2007).

Table 1- Qualitative composition per 100 g of whole wheat grains (Hebrard, 1996).

Constituents	Quantity	Constituents	Quantity
Water (g) Energy	13	Ca (mg)	35
(kj) Energy (kj)	1383	Mg (mg)	100
Carbohydrates (g)	331	P (mg)	390
Fat	63	Na (mg)	5
Protein (g, N*6.25)	2.5	K (mg)	-
Dietary fibre (g) Vit	14	Fe (mg) Vit B1	4.5
B2 (mg)	9.5	(mg) Vit PP (mg)	0.5
Vit E (mg)	0.09	Biotin	6
Folic acid (mg)	3		0.01
	0.04		

2. Impact of stress on wheat cultivation durum

2.1. Notion of stress

In the plant domain, the word stress has different meanings, but generally it is the change in a given physiological state caused by factors that tend to disturb the equilibrium of the plant system under study (Chinnusamy et al., 2006; Langridge et al., 2006). According to Maarouf and Raynaud (2007), stress is the set of physiological or pathological disturbances caused in the organism by biotic or abiotic agents. Tsimilli-Michael et al (1998), consider stress to be a relative meaning, with control as the reference state, they consider stress to be a deviation in control from a constraint. Naturally, plants must adapt to cope with biotic and abiotic stresses (Ishida et al., 2008).

2.2. The water deficit

Water deficit is the main environmental stress that severely limits agricultural productivity around the world (Boyer, 1982). It imposes constraints on growth and development of plants of agricultural interest, significantly reducing their productivity (Monnveux et al., 1986; Pereya et al., 2003; Kamoshita et al., 2008; Adda et al. 2013). There are many definitions of drought. In agriculture, it is defined as a deficit in rainfall that significantly reduces agricultural production compared with the normal for a given region (Mckay, 1985 in Bootsma et al., 1996). Passioura (2004) defines water deficit as the set of environmental circumstances in which plants experience a reduction in growth and production as a result of a deficient water supply. A water deficit occurs when the demand for water exceeds the quantity available over a certain period, or when its poor quality limits its use (Madhava Rao et al., 2006). The plant's water status progressively goes through three main phases, depending on how the drought develops (Tardieu, 2005). Under conditions of optimal water supply, transpiration and assimilation proceed in an equivalent manner. This situation continues until the amount of water absorbed is less than the plant's evaporative demand. Beyond this imbalance, a second phase begins, during which transpiration and assimilation are reduced relative to potential (Tardieu, 2003). The plant reacts by tending to restore the balance between these two physiological functions (absorption, transpiration) by externalising the appropriate mechanisms (Passioura, 2004). As the water deficit increases, the plant limits its gas exchanges by closing the stomata, which results in the cancellation of photosynthetic activities. Water exchanges with the atmosphere take place exclusively through the cuticle, so the plant begins to survive by consuming its reserves and losing all gains in dry matter production (Passioura, 2004).The constraints imposed by water deficit on photosynthesis relate in particular to a reduction in photosynthesising surface area, a reduction in stomatal conductance and an alteration in the photosynthetic apparatus (Sarda et al., 1992). However, these transformations are greatly influenced by the phases of development and the nature of the plant's organs. Generally speaking, the constraints generated by an increase in water deficit are imposed by a drop in relative water content and, consequently, an excessive reduction in the water potential of the plant's tissues. This situation results in the production of assimilates (Rebetzke et al., 2002) reserved for root growth and used for osmotic adjustment (Condon et al., 2004).

2.3. Effect of water deficit on durum wheat

Water deficit is one of the main factors limiting cereal productivity, particularly in winter. Its penalising effects are perceptible during all the processes involved in plant growth. It generates major changes affecting all the plant's morpho-physiological and biochemical traits, as well as levels of genetic expression (Mefti et al., 2000). The first consequences of water deficit on the plant are expressed through a reduction in photosynthesis following a drop in water potential which reaches -8.4 or even -20 bar. In such a situation, the transfer of assimilates ceases (Condon, 2002). According to Gharti-Chhetri and Lales (1990) in certain species of the Triticum genus, photosynthesis becomes zero when the leaf water potential reaches -27 bar. This reduction in activity is caused by stomatal closure, alteration of photosynthetic complexes and inhibition of Rubisco. These physiological consequences of water deficit affect the growth and development of roots, aerial parts and reproductive organs (Debaeke et al., 1996). On one scale, the depletion of sources of assimilate transfer to the developing ear alters its filling and yield development (Ricards, 1983; Benlaribi et al., 1991; Khaldoun et al., 1997). However, the effects of drought on the expression of the wheat plant depend on when it occurs in the development cycle (Debaeke et al., 1996). Drought occurring during the vegetative growth period mainly reduces the number of spikes per unit area. However, when it occurs during fertilisation, grain formation and grain filling, it reduces the number of grains per spike as well as its weight (Lawlor et al., 1981; Mogensen and Jensen, 1989; Triboï, 1990; Mogensen, 1991).According to Debaeke et al (1996), drought at the end of bolting, 10 to 15 days before heading, reduces the number of fertile flowers per spikelet. A deficient water supply after flowering, combined with high temperatures, leads to a reduction in the number of fertile flowers per spikelet. weight of 1000 grains by altering grain filling speed and filling time (Triboï, 1990).

2.4. Plant adaptation strategies to water deficit

2.4.1.Features phenological

Shirking is one of the strategies that consists of shortening the cycle of a cultivar in order to avoid late dry periods (Amigues et al., 2006). The use of short-cycle varieties is necessary in regions where drought occurs at the end of the cereal development cycle (Mekhlouf et al., 2006). Early heading is an important strategy for avoiding water deficit and heat stress at the end of the durum wheat cycle. Fischer and Maurer (1978) note that each day of earliness increases durum wheat yield.

2.4.2.Avoiding the constraints of drought

The avoidance strategy enables the plant to maintain a high water potential through particularly morphological changes involved in increasing water uptake capacity and reducing water loss rates. Such strategies prevent the declaration of stress at the level of the plant subjected to a water deficit (Amigues et al., 2006). Extensive root systems exploit the deep horizons of the soil, which remain wetter in drought conditions. By exploiting a greater volume of soil, the plant is able to meet its water requirements, maintain gas exchange and grow in drought conditions (Annerose, 1990). In addition, the reduction in leaf area, leading to a reduction in the evaporative surface, and leaf curling are effective morphological remodelling techniques for reducing transpiration (Cooper et al., 1983; Clarke, 1986; Benmahammed et al., 2008). However, this phenomenon is also accompanied by a reduction in photosynthesis, so it is only of interest in the presence of a severe water deficit (Slama et al., 2005). The reduction in plant height is often considered to be a mechanism for avoiding the stress generated by the water deficit (Annichiarico et al., 2005).

2.4.3. Tolerance of water deficit or physiological adaptations

The tolerance strategy is designed to lower the water potential of plant tissues in order to maintain cell turgidity (Monneveux and This, 1997; Sorrells et al., 2000). This process is externalized, mainly by osmotic adjustment (Zhang et al., 1999). According to Blum (1989), osmotic adjustment is achieved through an accumulation of polar solutes, mainly in the vacuole. This accumulation enables the plant to maintain its turgidity and avoid dehydration (Morgan et al., 1986). These solutes, often called osmoticums, are essentially organic acids, amino acids (proline, glycine-betaine), soluble sugars and certain inorganic constituents (Richards et al., 1997).

3. Structure and composition of wheat

3.1. Wheat grain structure durum

The wheat kernel is a dry, indehiscent fruit or caryopsis. Its morphology is characterised by an ellipsoid shape with a deep longitudinal groove of 1.5 to 2 mm, which extends along the entire length of the ventral surface. One end has hairs, while the other end has a germ. Some of these characteristics vary and depend on varieties, growing conditions and the position of the grain on the ear (Calderini et al., 2000, Evers and Millar, 2002 in Ferreira, 2011). At maturity, the wheat kernel is made up of three parts: the endosperm, the husks and the germ.

3.1.1. The Albumen

The albumen, which constitutes the reserve tissue, represents the largest part of the grain (80-85% of its dry weight). It is made up of two distinct tissues, the aleurone layer or protein base and the starchy albumen (kent et al, 1994 ; Barron et al., 2007).
Starch albumen cells are formed by starch grains enclosed in a protein matrix composed of prolamins, albumin and globulins (Zaddem, 2014). These nutrients are mobilised to support the growth of the embryonic axis at the start of germination (Fig.1).
The alcurone layer is made up of living cells and separates the starchy albumen from the envelopes (Fig.3). Along with the embryo, it is the only living tissue in the mature grain and enables it to develop during germination (Tab.2) (Stevenson et al., 2007).

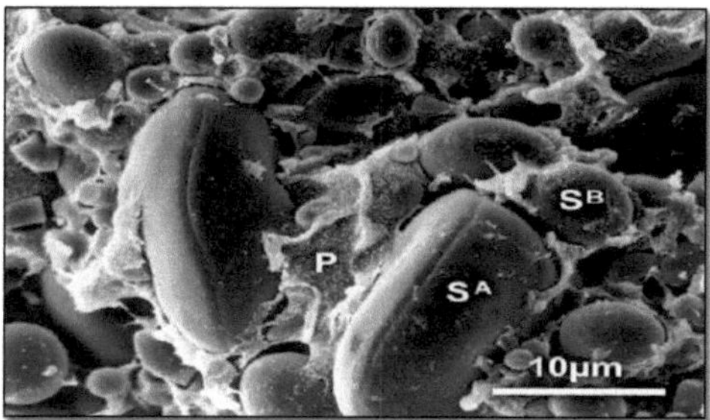

Figure 1- the part of Power plant image albumen obtained by scanning electron microscopy (SEM). Presence of two populations of morphologically different starch grains, small (SB) and large (SA), embedded in a protein matrix (Mills et al., 2005).

3.1.2. The germ or the embryo

It is made up of two parts, the embryonic axis consisting of the gemmule protected by the coleoptile and the radicle capped by the coleorrhiza and the scutellum (rudimentary cotyledon) (Surget and Barron, 2005).

3.1.3. The envelopes

They represent around 15% of the total mass of the grain. They are mainly composed of polysaccharides (arabinoxylans, xyloglucans and cellulose), phenolic acids, lignin, proteins and minerals (Godon and William, 1991; Surget and Barron, 2005). They vary in thickness and are made up of the outer pericarp, inner pericarp and testa from the periphery to the centre (Fig. 2).

Table 2- Schematic representation of the distribution of compounds of nutritional interest in wheat grain (adapted from Hemery et al, 2007).

Peripheral tissues	Pericarp	Testa	Aleurone	Starchy albumen	Germ
Proteins	-	-	**	**	*
Lipids	-	*	*	-	***
Starch	-	-	-	***	-
Insoluble fibres	***	***	**	*	**
Fibres soluble	-	-	*	**	*
B vitamins	-	-	***	-	**
Vitamin E	-	-	*	-	**
Minerals	*	*	***	-	**
Plant sterols	-	*	**	-	***
Phyto-oestrogen Alkylresorcinols	-	-	***	-	-
	-	***	-	-	-

*** - - - *, **, *** indicate whether a compound is present, concentrated, or highly concentrated in a tissue.

- : indicates that a compound is not present in the tissue or is present in very low concentrations.

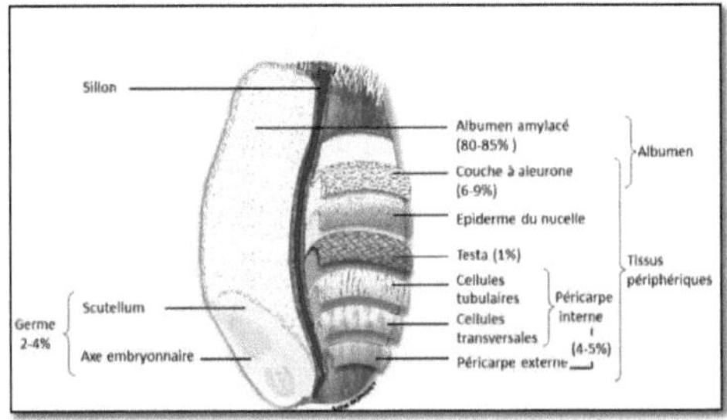

Figure 2- Longitudinal and transverse section of a wheat grain (Surget and Barron, 2005).

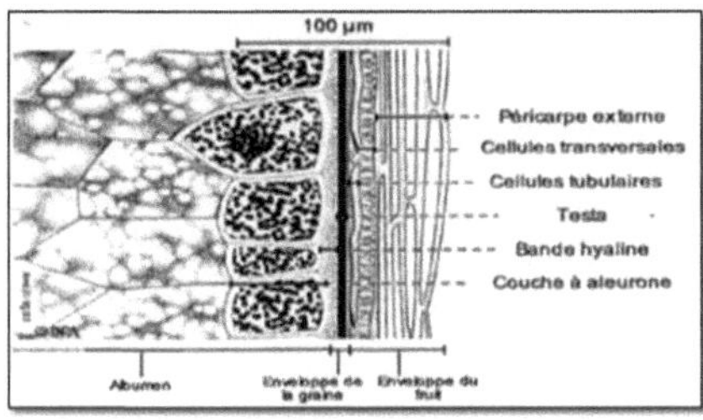

Figure 3- Cellular layer of wheat grain (Surget and Barron, 2005).

3.2. Composition of wheat grains durum

At maturity, durum wheat grain is essentially composed of starch (70%) and protein (10%-18%). The starch is concentrated in the starchy albumen and the proteins are mainly found in the germ and the peripheral aleurone layer (Feillet, 2000).

3.2.1. Starch

Starch is deposited in the amyloplasts in a semi-crystalline form that is insoluble in water. In cells, it is present in the form of grains visible under the microscope in the amyloplasts (Bayuelo-Jimenez et al., 2002). It consists of the coexistence of two polymers of D-glucopyranose, amylose and amylopectin (Fig. 4). These two compounds differ in the degree of polymerisation and the number of branches (Sestili et al., 2010). In cereals, in addition to the two main constituents, it also contains lysophospholipids and free fatty acids (Buléon et al., 1998).The starch in wheat grain consists mainly of amylopectin, with an average content of around 75%. The amylose fraction is lower, at around 25%. Amylose is made up of long linear chains (10^4 units) of D- glucopyranose polymer (Hoseney, 1994; Debiton, 2011). Amylose has the ability to create helical structures around free fatty acids. Amylopectin, a polymer of the monomer D-glucopyranose, consists of a linear column formed by α (1-4) osidic bonds, from which lateral branches are formed by α (1-6) bonds (Colonna and Buléon, 1992).

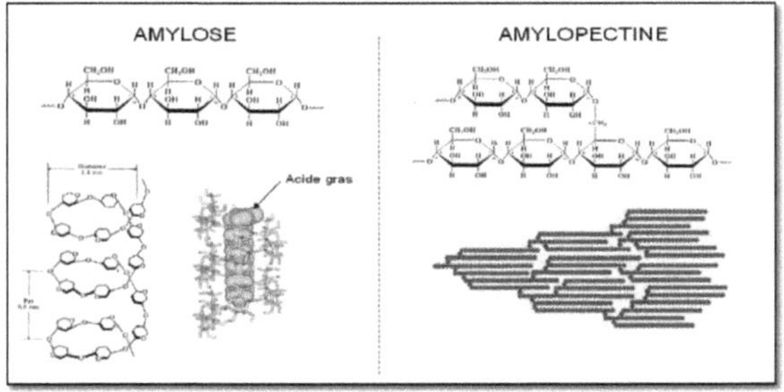

Figure 4- Structure of amylose and amylopectin (Debiton, 2010).

The starch grains stored in the wheat kernel are of three types (Fig. 5) which vary according to their cheeses and sizes. Starch is stored in the form of granules, which are distinguished by their size and shape (Stoddard, 1999; Wilson et al., 2006; Stamova et al., 2009). Type A grains have a diameter greater than 16 µm and a lenticular or elliptical shape. Type B grains have an average diameter ranging from 5.3 to 15.9 µm and are spherical in shape (Morrison and Gadan, 1987). Finally, type C grains have diameters of less than 5.3µm (Bechtel et al., 1990).

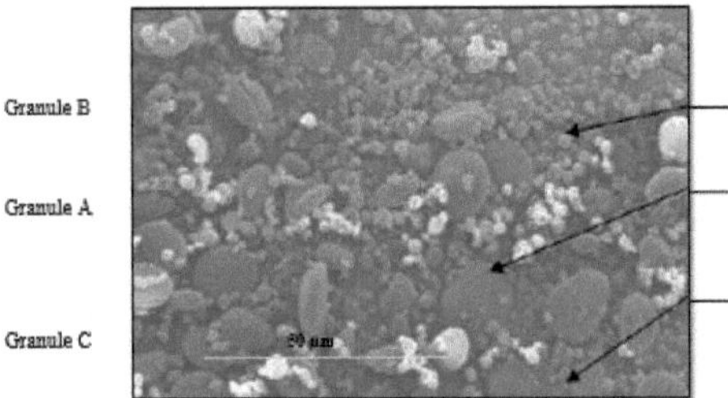

Figure 5- Wheat starch grains (Debiton, 2010).

3.2.2. The proteins

The protein fraction of durum wheat grain, which represents around 12% of the dry weight, is located mainly in the embryo and the aleurone layer (Battais et al., 2007; Saulnier, 2012). However, this content can vary according to environmental factors and the intrinsic characteristics of cultivars (Hernandez et al., 2004). Depending on their solubility medium, wheat grain proteins can be divided into four groups (Battais et al., 2007; Ayad et al., 2010; Saulnier, 2012): albumins, globulins, prolamins and glutenins. These types of protein can be divided into two main categories, soluble proteins and reserve proteins (Fig. 6).

3.2.2.1. Soluble proteins (metabolic proteins)

This category includes albumins, globulins, enzymes, membrane proteins, regulatory proteins and cell organelle proteins (Belitz and Grosh, 1987; Singh and Skerritt, 2001; Vensel et al., 2005; Sramkova et al., 2009).

3.2.2.2. Proteins from reserves

The largest fraction of proteins in this category accumulate in the albumen, making them reserve proteins (Hernandez et al., 2004). These proteins are also called the prolamine category because of their high content of proline residues. and glutamine (Shewry et al., 1986; Saulnier et al., 2012). The reserve proteins contain two types, gliadins and glutenins, each of which can be distinguished into several subtypes (Fig.6).

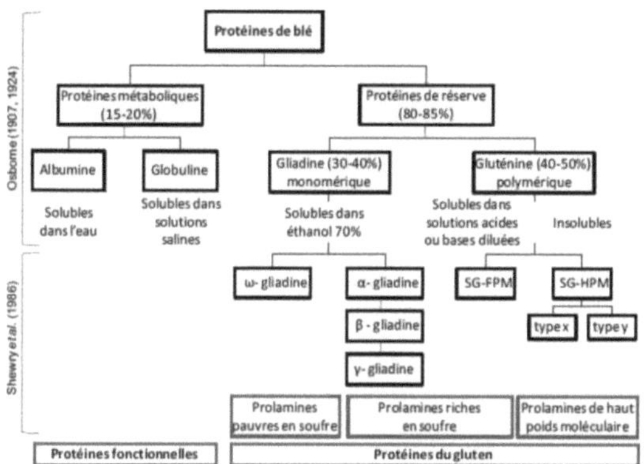

Figure 6- Classification of wheat proteins. Comparison of the classifications of Osborne (1907) and Shewry et al. (1986).

4. Grain filling in durum wheat

The filling phase, which lasts around 45 days, is marked by grain enlargement (Battinger, 2002), accumulation of reserves and gradual dehydration, which marks the end of maturation (Soltner, 1988). The grain's starch and protein reserves are accumulated in the albumen, which consists of the peripheral aleurone layer and starch-rich inner cells (Barron et al., 2012). Starch and protein filling begins at the end of the organisation of the reserve tissue, marked by cellular compartmentalisation of the albumen. Various genetic and environmental factors affect t h e synthesis and accumulation of starch and proteins. proteins in grains (Dupont and Altenbach, 2003). Panozzo and Eagles (1999) have shown that the rates and durations of starch and protein accumulation are considered to be independent events, controlled by distinct mechanisms. These factors are distinguished by the genetic nature of the cultivars and climatic parameters, particularly temperature and rainfall (Yin et al., 2009).

Starch synthesis comes mainly from photosynthesis during grain filling, while nitrogen accumulation in the grain comes mainly from the nitrogen assimilated and stored during the vegetative phase (Wiegand and Cuellar, 1981; Bogard, 2011). There are three main phases in grain development (Olsen, 2001).

Milky grain phase: characterised by cell multiplication and the organisation of the reserve parenchyma constituting the albumen. During this period, the water content of the cells increases, accompanied by a weak constitution of the starch grains and the deposition of proteins (Carceller and Aussenac, 1999; Dupont and Altenbach, 2003). At this stage, the grain envelopes differentiate. It should be noted that the speed and intensity of reserve accumulation depend closely on the state of cell hydration and the organisation of the albumen, depending on the number and size of the cells in it (Egli, 1998).

Paste phase: this is the main stage of grain filling, during which the reserves migrate and organise themselves within the grain. This phase is characterised by an increase in grain volume induced by intense polymerisation of simple sugars into starch and protein deposition (Sofield et al., 1977; Singh and Jenner, 1982; Jenner et al., 1991, Panozzo and Eagles, 1999; Shewry et al., 2009). It is during this phase that starch and protein levels stabilise (Calderini et al., 2000) and the quantity of water absorbed is equal to that evapotranspired (Gooding et al., 2003).This period is known as the hydric plateau phase, when grain moisture levels are around 40-45%.

Physiological maturity phase: this is known as the grain desiccation or physiological maturity phase (Boufenar and Zaghouane, 2006). It is marked by the cessation of the migration of reserves into the grain and a reduction in its moisture content (15% to 16%), and its final weight is reached (Gate, 1995; Wardlaw, 2002).

4.1. Sources of fill for grain

During grain filling, assimilates are supplied by two main sources, current photosynthesis in the vegetative organs and remobilisation of reserves during the post-anthesis period (Blum et al., 1991; Bahlouli et al., 2008). The proportion of each source in this process depends on the quality of the plant's water supply, where water is a determining factor in the longevity of the vegetative organs and the circulation of sap (Amigue et al., 2006).

4.1.1. Photosynthesis

The post-anthesis photosynthetic activity of the vegetative organs, mainly those located above the last node, is an important source of grain filling (Fellah et al., 2002; Bahlouli et al., 2008). Thus, a more developed stem vegetative apparatus, particularly the leaf system, provides consistency in the photosynthetic apparatus and greater efficiency in the production of assimilates essential for grain filling (Belkharchouche et al., 2009). However, this activity depends on the longevity of the organs concerned. In fact, this longevity, which determines the duration of activity of these organs, is conditioned by the plant's water nutrition. The declaration of drought during this phase reduces this activity and limits the involvement of this source in grain filling and formation. The main organs responsible for this activity are the foliage, particularly the last leaf, the neck of the ear, the beard and the constituents of the ear (envelopes, the developing seed).

4.1.1.1. The cob

Through its photosynthetic activity, the ear plays an essential role in the assimilation of assimilates into the grain during its formation and filling (Bort et al., 1994). The contribution of this organ to the overall photosynthetic yield of the plant can represent a significant proportion varying between 13 and 76% (Biscope et al., 1975), depending on the crop. climatic conditions. This contribution increases more in situations of terminal water deficit, where the ear is the main organ responsible for yield development in durum wheat (Gate et al., 1993; Bammoun, 1997). In drought conditions, photosynthesis in the ear plays a relatively greater role in grain filling than the flag leaf.However, the ear contributes to the increase in water loss by the presence of barbs dotted with lenticels as well as tomatoes, which gives it an epidermal conductance (Ali Dib et al., 1992). Nevertheless, according to Febrero et al (1990), cobs with short barbs help to limit this water loss.

4.1.1.2. The beards

The presence of beards in cereals increases the possibility of water use and dry matter production during the ripening phase (Nemmar, 1980). They help to improve the overall photosynthetic yield of the plant, and according to Bort et al (1994) and Fokar et al (1998), photosynthesis in bearded ears is less sensitive to the inhibitory action of high temperatures than in hairless ears. According to Hurd (1974), when the last leaf senesces, the contribution of the beard to yield reaches 7%, giving bearded wheat an advantage (Hadjichristodoulou, 1985).

4.1.1.3 The l'épi pass

The role of this part of the stem is consolidated by its current photosynthetic activity during the grain filling phase and the assimilates stored at this level are likely to be transported to the developing grain, essentially in situations of terminal water deficit (Gate et al., 1992). Bahlouli et al (2008) report that assimilates from this organ are higher under unfavourable climatic conditions than under favourable conditions. However, a positive correlation has been noted between the length of the spike neck and plant height (El- Hakimi, 1995).

4.1.1.4. The last leaf

The contribution of the last leaf to the final weight of the grain depends on the climatic conditions that affect its surface area, longevity and photosynthetic activity. Some studies (Planchon, 1976; Abbad et al., 2004; Tambussi et al., 2005, 2007; Lopes et al., 2006; Khaliq et al., 2008) show that the contribution of the last leaf is important in determining the final yield of durum wheat.According to Clarke (1989) and Monneveux (1991), the lifespan of the last leaf, estimated by the change in its green surface, provides information on the level of functioning of the photosynthetic apparatus in the presence of a water deficit. According to Auriau (1978), the longevity and photosynthetic activity of the last leaf greatly influence grain yield in cereals.

4.1.2. Remobilising reserves

The vegetative phase of a cereal crop, during which the various vegetative organs are formed, has a major impact on yield performance. It is in these organs that the nitrogen and hydrocarbon reserve tissues are organised (Bogard, 2011). During the reproductive period, the translocation of these assimilates plays an effective role in filling the grain and consequently in determining its final weight and the quality of its reserves (Belkharchouche et al., 2009). The relative importance of remobilised assimilates in filling the grain depends on the quantity of dry matter accumulated at the heading stage (Bahlouli et al., 2008; Belkharchouche et al., 2009). The involvement of remobilisation in grain formation depends on the duration of the anthesis-maturity phase. According to Blum (1998), early growth or avoidance of the terminal water deficit implies little use of its reserves, whereas late growth is characterised by major remobilisation of these reserves from the stubble. However, in drought conditions, high-straw cultivars, which are generally late, are advantageous because of the high quantity of assimilates stored and remobilised during the grain filling phase, thus improving grain weight and quality (Simane et al., 1993; Bahlouli et al., 2008). The tall straw character is also associated with more pronounced root elongation, giving the plant greater water extraction capacity (Bagga et al., 1970).

MATERIALS AND METHODS

1. Objective of the work

Grain filling is a key stage in the development of durum wheat yield. This phase is largely conditioned by the intrinsic properties of the cultivars grown and the environmental effects that occur during the process. The interaction between these two factors promises to be extremely important. Grain filling and grain quality depend on the availability of photo-assimilates at this stage, which come from current photosynthesis and the remobilisation of reserves stored during the plant's vegetative period. The main objective of the work carried out in the thesis is to elucidate the intervention of the two processes involved in grain filling under water deficit conditions.The interaction between genetic variability in durum wheat and variations in water supply was also studied. The involvement of the various plant organs responsible for current photosynthesis and remobilisation of reserves in grain filling and quality development has been elucidated. These studies were carried out on different genotypes subjected to different water-deficient regimes under field and controlled conditions.

2. Field trials

The field trials were set up at the experimental farm of the Ibn Khaldoun University in Tiaret (34°04' and 35°11' north latitude and between 1°33' and 1°53' east longitude; at an altitude of 1031m), during three consecutive seasons (2012/2013, 2013/2014, 2014-2015). The system adopted comprises two treatments, one under irrigated conditions and the other under rainfed conditions. In each treatment, the genotypes were arranged in random blocks with three replications for each (Tab. 3). The sowing rate used was 21 plants per line, 1 m long and 20 cm apart. During the trials, the necessary manual weeding was carried out. Three applications of ACTIVEG commercial nutrient solution were made for each trial.With regard to the irrigated treatment, make-up irrigation by immersion was carried out every week during periods of agronomic drought over the course of the three irrigation campaigns.

Table 3- Experimental set-up

36	27	28	39	30		36	27	28	39	30
31	38	33	34	40		31	38	33	34	40
26	37	32	29	35		26	37	32	29	35
41	52	43	44	45		41	52	43	44	45
46	47	68	49	50		46	47	68	49	50
51	42	53	54	59		51	42	53	54	59
56	57	10	73	69		56	57	10	73	69
61	62	63	64	72		61	62	63	64	72
66	67	48	60	70		66	67	48	60	70
71	65	59	74	75		71	65	59	74	75
76	88	78	79	80		76	88	78	79	80
81	94	93	84	85		81	94	93	84	85
86	92	77	83	95		86	92	77	83	95
91	87	89	82	90		91	87	89	82	90
99	104	5	8	100		99	104	5	8	100
96	109	97	105	98		96	109	97	105	98
101	9	102	118	115		101	9	102	118	115
1	111	2	4	112		1	111	2	4	112
58	106	116	110	103		58	106	116	110	103
114	3	6	7	119		114	3	6	7	119
108	120	117	107	113		108	120	117	107	113
11	16	22	24	14		11	16	22	24	14
13	20	15	18	12		13	20	15	18	12
23	17	21	25	19		23	17	21	25	19

Treatment under rainfed conditions   Treatment under irrigated conditions

2.1. Plant material

The plant material consists of 120 genotypes of durum wheat (Triticum durum Desf.) of different origins. The collection includes varieties grown in Algeria, including those selected by local populations. The plant material was supplied by the Institut technique des grandes cultures (ITGCT) in Tiaret. The list of genotypes used and their origins is presented in the appendix (Appendix Tab.1). The selection criteria for the genotypes used were based in particular on their origin, phenological characteristics, morphological features and degree of tolerance to water deficit.

2.2. Climatic conditions of the campaigns

Climatic data for the experimental period was provided by the Tiaret weather station. The various climatic parameters during these periods were highly variable, and their details will be presented later and separately for the different campaigns.

2.2.1. Campaign 2012/2013

Annual rainfall during this season was 632.8 mm, recorded mainly between September 2012 and June 2013 (Fig. 7). November, January and March were the wettest months, with 322.6 mm.

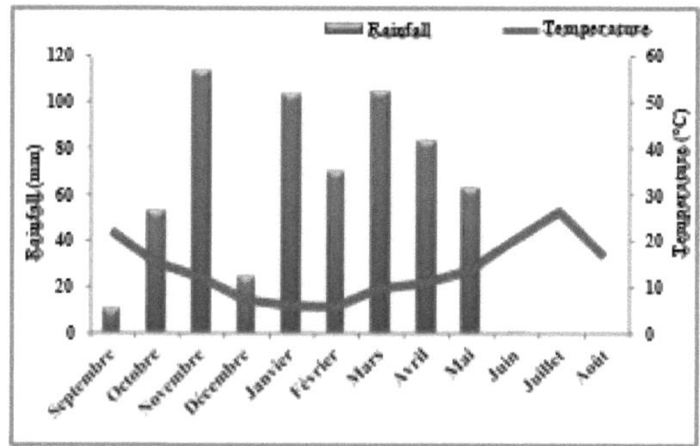

Figure 7- Umbrothermal diagram for the 2012/2013 campaign.

The average monthly temperatures recorded during this campaign varied. The highest average temperatures were recorded in June and July 2013. The lowest average temperatures were recorded between December and March.

2.2.2. Campaign 2013/2014

The total rainfall recorded during the 2013-2014 season was 425.5mm, which represents a deficit of 207.3mm compared with the first season of experimentation (Fig. 8). December, January, February and March were the wettest months of the year, with a cumulative rainfall of 331.5mm. Rainfall was lowest in April (0.7mm) and May (4.8mm), which coincide with the grain filling phase. Average maximum and minimum temperatures were recorded in June (33.46°C) and December (0.7°C) respectively.

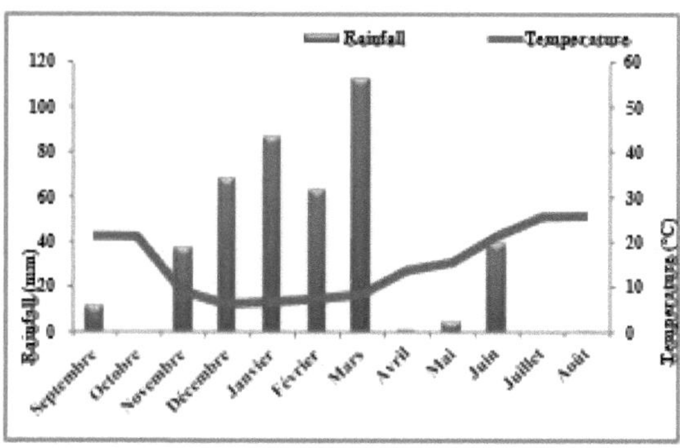

Figure 8- Umbrothermal diagram for the 2013/2014 campaign.

2.2.3. Campaign 2014/2015

The 2014/2015 season was characterised by rainfall of 445 mm (Fig.9), most of which was received during the period between September, December and February (291 mm). This season was characterised by an agronomic drought spread over a period of 5 months (March-July).

Average temperatures ranged from a low of 4.91°C (February) to a high of 21.07°C (June).

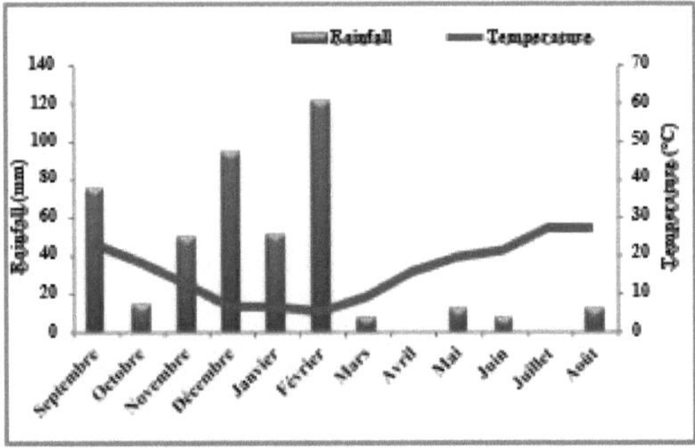

Figure 9- Umbrothermal diagram for the 2014/2015 campaign.

2.3. Setting up the filling system

In order to assess the contribution of current photosynthesis by the various plant organs and the remobilisation of organs in grain filling, a system was adopted from the anthesis phase onwards. To this end, the whole plant was shaded by wrapping it in aluminium foil to determine the contribution of remobilisation of reserves to grain filling. Assessment of current photosynthesis was carried out by shading the neck of the ear and the ear with aluminium foil and by excising the last leaf and the barb of the ear (Fig. 10). Control plants were maintained without any manipulation.

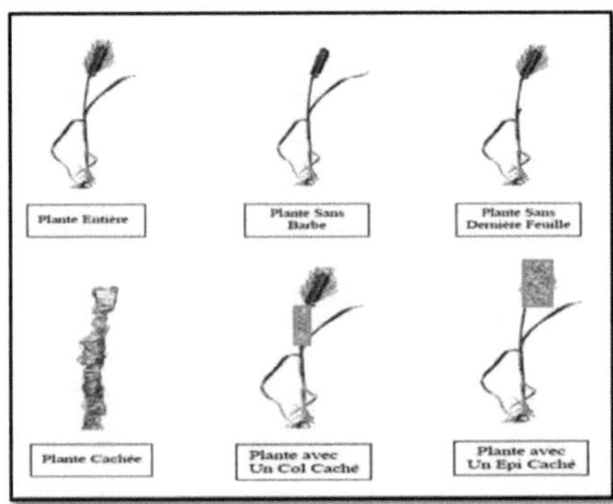

Figure 10- Experimental set-up for field trials

2.3.1.Measurements

2.3.1.1. Morphological characterisation of the genotypes used

When the tested genotypes reached physiological maturity, morphological measurements were carried out on the plants:
- Plant height (cm)

- Length of last knot (cm)

- Ear neck length (cm)

- Ear length (cm)

At the same stage, the weight of a thousand grains was determined in the control plants and those subjected to the filling device. The rate of contribution of the organs involved in this process and the remobilisation of reserves was determined using the following method:

- The contribution of current photosynthesis by vegetative organs

$$\text{Taux de la contribution} = \left[\frac{\text{(PMG plante Témoin - PMG plante avec organe ombré ou excisé)}}{\text{PMG plante Témoin}}\right] \times 100$$

- La contribution de la remobilisation des réserves

$$\text{Taux de la contribution} = \left[\frac{\text{PMG plante Témoin - PMG plante entièrement cabée}}{\text{PMG plante Témoin}}\right] \times 100$$

- The contribution of remobilising reserves

3. Tests conducted under controlled conditions

The greenhouse experiments were carried out to control the water factor separately from other climatic parameters. The plants were subjected to different levels of water supply. This made it possible to distinguish the effect of water regimes on grain filling and grain quality.Two trials were carried out under these conditions. In the first trial, the contribution of current photosynthesis and remobilisation of reserves to grain filling was studied. The second trial assessed the kinetics of grain filling.

3.1. Plant material used

The plant material consists of five durum wheat (Triticum durum Desf.) genotypes of different origins and drought tolerance. The collection includes two genotypes from local populations, Ouad zenati and Langlois. The other three genotypes, Waha, Acsad 1361 and Mexicali 75, were introduced from ICRDA, ACSAD and CIMMYT respectively. The main agronomic characteristics of the genotypes are given in Table 4.

Table 4- Main agronomic characteristics of the genotypes used.

Genotype code	Genotype name	Origin	Drought tolerance	Vegetative cycle
01	WAHA	ICARDA	High	Early
02	ACSAD1361	ACSAD	Low	Semi-late
03	MEXICAL 75	CIMMYT	Medium	Early
04	OUAD ZENATI	LOCALE	Medium	Tardif
05	LANGLOIS	LOCALE	High	Tardif

3.2. Conditions for conducting tests

The experiment was carried out in an automatic greenhouse at the Faculty of Natural and Life Sciences at Ibn Khaldoun University in Tiaret. Disinfected and pre-germinated seeds were sown in PVC cylinders 120 cm long and 20 cm in diameter, filled with a homogeneous substrate made up of sand, soil and compost in proportions of 8:3:1 respectively. The temperature in the greenhouse was maintained at approximately 25°C (daytime) and 15°C (night-time) and the relative humidity at 70%.The cylinders were arranged in three batches with different water treatments (Fig. 11 and 12). From installation to the anthesis stage, all the cylinders were brought up to field capacity by a daily addition of 300 ml of water. The irrigation water was replaced four times by a commercial nutrient solution of the ACTIVEG type. After anthesis, the irrigation regimes were modified. For the control treatment, the cylinders were maintained at field capacity (100%CC) until maturity, while the other two batches were run under water deficit, evaluated at 60% of field capacity (60%CC) and 30% of field capacity (30%CC). For each water treatment, each genotype was repeated five times, giving a total of 25 cylinders per water treatment and 75 cylinders for the three water regimes.

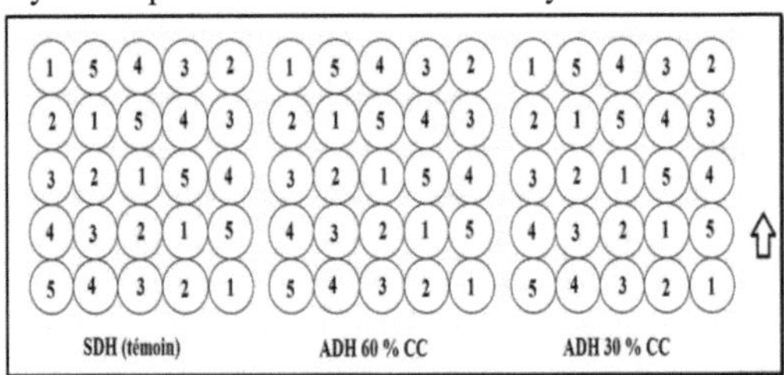

Figure 11- Schematic diagram of the experimental set-up.

Figure 12- Vegetation in place, bolting stage.

To assess the contribution of the different parts of the plant to grain formation, a system was adopted (Fig.13). For each water treatment and for all the replicates, the plants were divided into six treatments. Control plants, plants without the last leaf, plants without a beard, plants with a hidden ear, plants with a hidden collar, fully hidden plants and plants without all the leaf stages. The various organs were removed using a pair of scissors. The various organs were hidden using aluminium foil. Very fine perforations were made in the aluminium foil to improve air circulation and gas exchange with the environment. Figure 13 summarises the treatments applied to the plants.

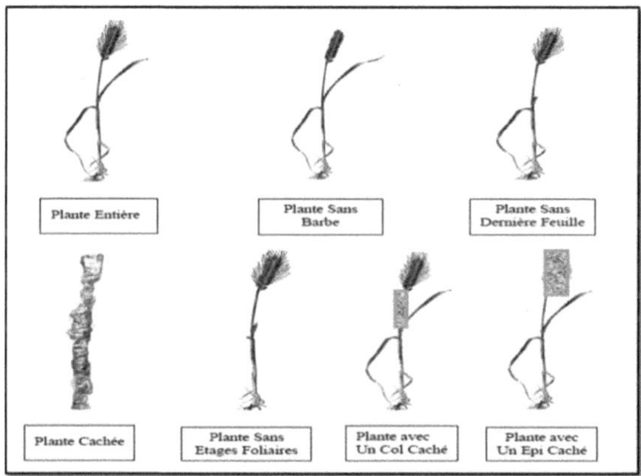

Figure 13- Diagram of the filling system by shading and excision of organs.

3.3. Measurements

3.3.1. Trial 1: Contribution of organs to the filling of grain

At physiological maturity, a morphological characterisation of the genotypes tested was carried out. The following measurements were taken:

- Plant height (cm)

- Length of last knot (cm)

- Ear neck length (cm)

- Ear length (cm)

- The length of the beard of the ears (cm)

- Leaf area determined by an electronic planimeter (cm^2)

The thousand kernel weight was determined for all plants in the three treatments, the controls and the filling system. From the values of this yield component, the rates of contribution to grain filling by current photosynthesis and remobilisation of reserves were determined according to the following formulae:

- The contribution of current photosynthesis by vegetative organs

$$\text{Taux de la contribution} = \left[\frac{\text{(PMG plante Témoin -PMG plante avec organe ombré ou excisé)}}{\text{PMG plante Témoin}}\right] \times 100$$

- La contribution de la remobilisation des réserves

$$\text{Taux de la contribution} = \left[\frac{\text{PMG plante Témoin -PMG plante entièrement cahée}}{\text{PMG plante Témoin}}\right] \times 100$$

4. Analysis of results

The results of the various tests were subjected to statistical analysis (ANOVA) using STATISTICA Version 8.0.

RESULTS

1. Field trials -

1.1. Characterisation and effects of the water regime on the main morphological parameters of the genotypes tested over three years of experimentation.

The grain filling process in durum wheat is essentially based on reserves developed during the grain formation phase. These processes are linked to the morphological behaviour of the plant and the quality of the environmental factors in which these characteristics are expressed. It should be noted in this context that water supply and the variability of the plant material are the main factors in determining the quality and final weight of the grain. This part of the work concerns the identification of the aerial vegetative organs responsible for the development and translocation of reserves during the grain filling phase. The first stage involves morphological characterisation of the genotypes used and an assessment of the effects of variations in water supply on the expression of these characteristics. The second part will assess the contribution of these organs to grain filling.Analysis of the results reported in Table 05 for the study campaigns (2012-2013, 2013-2014 and 2014-2015) shows that the genotypes tested have very distinct morphological characteristics and that their expression is strongly influenced by the water regimes adopted (p<0.001). Indeed, the absence of irrigation was accompanied by a clear reduction in the magnitudes of the morphological traits concerned by this study. The effects imposed by the interaction of the nature of the genotypes, the management campaigns and the water regimes applied proved to be significant on the expression of these traits, thus heralding a genotypic distinction with respect to the environmental factors for their development (p<0.001). These parameters concern the height of the plant, the length of the last node, the length of the neck of the ear and the length of the ear. The weight of a thousand grains was determined and used to estimate the contribution of the organs presented later in this part of the work.

Table 5- Effects of genotypes, water conditions, year of trials and their interaction on the expression of the morphological traits selected.

	HP	LDN	LCE	LE
Genotypes	70,299***	51,914***	18,688***	3,0737***
Water situation	726,440***	248,9285***	22,991***	0.31701 ns
Year	754,5709***	311,4667***	59,997***	11,2308***
Genotypes*Water conditions	15,259***	24,0181***	8,8059***	1,91804***
Genotypes*Year	24,144***	10,0287***	3,638***	1,5982***
Water situation*Year	6,6942***	1.717 ns	1.9648 ns	7,5031***
Genotypes*Water status*Year	7,747***	6,0579***	2,540***	1.1372 ns

HP = Height of the plant, LDE = Length of the last internode, LCE = Length of the neck of the ear, LE = Length of the ear, ***significant effect at p ≤ 0.001

27

The results obtained from trait measurements during the 2012-2013 study campaign are highly variable among the genotypes tested and between the water regimes adopted and their interaction (Tab. 06).The results obtained for plant height (Appendix Tab.2) show that the values obtained for the irrigated treatment range from 111.00cm±2.64 (Langlois) to 49.33cm±0.59 (CNDO/VEE//7*PLATA_8/3/GUANAY/10/ PLATA_10 /6/MQUE/4/...).In the treatment carried out under rain-fed conditions, the values of this trait were between OUED ZENATI (85.50cm±1.04) and Guanay/3/Stot//Altar84/ADL (38.16cm±3.03). It can be seen in this context that the absence of irrigation led to variable reductions in the magnitudes of this trait, which fell within the 58% and 2% range recorded respectively in the ACSAD1351 and OUED ZENATI genotypes.Ear neck length also varied between genotypes within and between the two water treatments. In fact, in the irrigated batch, lengths varied between a maximum of 33.33cm±0.72 and a minimum of 9.43cm±1.32 recorded in order by Gloire de Mongolfier and KOFA/3/SOMAT_3/PHAX_1//TILO_1/LOTUS_4. For this character, the effect of the water deficit was remarkable where the reduction in its length reached a value of 26.73cm±0.83 (OUED ZENATI). On the other hand, the LANGLOIS genotype showed the lowest sensitivity to the development of this length, with a reduction of 0.43±0.12cm.Regarding the length of the last node, the same effects of the variation in water supply were observed. In the irrigated lot, the lengths ranged from 21.06 cm±2.21 (ACSAD1359) to 54.01 cm±2.12 (POLONICUM). The average reducing effect of water deficit on its expression was 10%, ranging from 1% (ACSAD1367) to 32% (Ofanto/Boussalem 1s).The average results for ear length show that the highest values are recorded for the genotypes in the irrigated system. At this level, the Langlois genotype stands out with a high length of 7.21cm±0.28, while the WAHA genotype has the lowest value at 3.33cm±0.44. Under rainy conditions, all the genotypes showed a reduction in this parameter with an average value of 11%. The highest reduction was recorded by the WAHA genotype (42%), while the lowest was recorded by the ACSAD1305 genotype (1%).

Table 6- Effects of genotype, water regime and their interaction on the expression of morphological parameters measured during 2012-2013.

	HP	LDE	LCE	LE
Genotypes	6,3189***	2,4500***	3,8050***	2,3410***
Water situation	55,7478***	5.2642 ns	6,5108**	16,525***
Genotypes*Water conditions	4,5668***	1,5269***	2,7759***	1,6718***

HP = Height of the plant, LDE = Length of the last node, LCE = Length of the neck of the ear, LE = Length of the ear; ***significant effect at p≤ 0.001

1.1.2.2013 campaign- 2014

During this year of experimentation (2013/2014), the effects of the nature of the genotypes grown and the water supply adopted were significant on the results of the parameters measured (Tab. 7). Plant height, the length of the last internode, the length of the neck of the ear and the length of the ear varied greatly among the genotypes tested within and between the two water regimes applied. The interaction between genotype and water regime also exerted an important influence on their expression. This indicates a genotypic distinction with respect to variations in water supply levels for the development of these characteristics.
The average results recorded (Appendix Tab.3) of plant height measurements show that its values in the irrigated lot oscillate between the extreme values of 46.3cm±0.49 (Sooty-9/Rascon-37/4/Storlom/3/Rascon37/...) and 100.67cm±6.17 (OUED ZENATI). The Gloire de Mongolfier genotype also recorded a height greater than 100cm with 100.23cm±0.96. The absence of irrigation resulted in clear reductions in the height of the genotypes tested. The average values obtained from measurements taken on this parameter ranged from 30.33cm±0.88,recorded by the Hubei//Sooty_9/Rascon_37/3/2*Sooty_9/Rascon_37/4/Sooty_9/Rascon_37 genotype, to 83cm recorded jointly by the DUILIO and OUED ZENATI genotypes.
The mean values obtained from measurements of the length of the last internode in irrigated plants show that they fall within the range delimited by extremes of 12.33cm±0.20 (ACSAD1339) and 42.03cm±1.15 (ACSAD1359). For the same line and under rainy conditions, the values obtained were reduced and ranged from 11.56cm±1 (KOFA/3/SOMAT_3 /PHAX_1// TILO_1/LOTUS_4) to40.83cm±8.92 (LLARETA INIA/4/SKEST//HUI/TUB/3/SILVER/5 LHNKE/RASCON//...)
The values obtained from measurements of ear neck length also varied among the genotypes tested. In the absence of irrigation, more or less significant reductions in the length of this parameter were observed. In the irrigated batch, the ear neck length values obtained varied between 6.16cm±0.44 (WAHA) and 23.37cm±0.63 (ACSAD1339). For rainfed plants, the lengths of this characteristic are reduced and fluctuate between extreme values of 2.4cm±1.3 and 21.7cm±0.85 recorded respectively by the genotypes LANGLOIS and Rascon/Dukem/3/THB/CEP7780//2*Musk-4/5/....

Variations in the results obtained from measurements of ear length appear to be more sensitive to the genotypic effect than those imposed by the water regimes applied. In the irrigated treatment, the values recorded ranged from 3.47cm±0.29 to 7.1cm±0.1 for the genotypes CNDO/VEE//CELTA/3/PATA_2/6/ ARAM_7/ CREX/ALLA/5/ENTE/... and Ajaia-16//Hora/jro/3/Gan/4/Zar/5/Malmuk-l..... In the absence of irrigation, under rain-fed conditions, the GUEMGOUM R' KHEM and YAVAROS 79 genotypes stand out from the rest, with extreme values of 3.8cm±0.3 and 7.43cm±0.72 respectively.

Table 7- Effects of genotype, water regime and their interaction on the expression of morphological parameters measured during the 2013-2014 season.

	HP	DN	CE	PPE	PMG
Genotypes	5.56***	5.42***	5.64***	3.63***	5.48***
Water situation	219.02***	66.26***	67.9***	6.14*	1121.21***
Genotypes*Water conditions	2.91***	2.92***	2.76***	2.38***	2.96***

HP = Height of the plant, LDE = Length of the last internode, LCE = Length of the neck of the ear, LE = Length of the ear, ***= significant effect at p< 0.001

1.1.3.2014 campaign - 2015

Analysis of the results obtained during this campaign (Tab. 8) shows that all the morphological parameters selected (plant height, length of the last internode, length of the neck of the ear, length of the ear) are greatly influenced by variations in the nature of the genotypes and the water regime adopted. The interaction of these two factors also leads to significant variations in the results obtained, indicating a genotypic distinction with regard to changes in water supply, for the expression of these morphological characteristics.With regard to plant height, the average results obtained in the irrigated lot (Appendix Tab.4) indicate that its values fluctuate between extremes of 49.33cm±0.6 and 111cm±2.65 held respectively by genotypes CNDO/VEE//7*PLATA_8/3/ GUANAY /10/PLATA_10/6/MQUE/4/... and LANGLOIS.For this same parameter and under rainfed conditions, the values obtained were within the range of 30.33cm±13.34 and 85.5cm±1.04 expressed in order by genotypes 35 and 114. These results indicate that the absence of irrigation, accompanied by the onset of a water deficit, caused a reduction in plant height in most of the genotypes tested, with an average rate of 16%. It should be noted that this rate varied considerably between genotypes.There were also noticeable variations in the length of the last internode, due to the nature of the genotypes and the water regimes adopted. In the irrigated treatment, they ranged from 19.83cm±2.68 to 45.07±8.69, in the Artico/Ajaia-3//Hualita//3/2*Somat-4/Inter-8 and ACSAD1357 genotypes respectively. Under rainfed conditions, these extreme values are held by the BONTE and Rascon_22/Rascon_21//Mojo_2/3/Guanay/4/Rcol/5/Sora/2*plata_12 genotypes. /Somat_3 with 13cm±2.52 and 54.23cm±1.39 respectively. The values of the length of the neck of the ear recorded in the irrigated lot are between the limits of 7.33cm±0.33 and 23.33cm±5.81 respectively expressed by the genotypes Stot//Altor84/ALD/3/Green-18/focha-l/... and Hubei//Sooty_9/Rascon_37/3/2. *Sooty_9/Rascon_37/4/Sooty_9/Rascon_37. On the other hand, in the batch grown under rainfed conditions, ear neck lengths varied between 0.43cm±0.12 (LANGLOIS) and 33.33cm±0.73 (GLOIRE DE MONGOLFIER). It can be seen that the ear neck is the morphological parameter that is most sensitive to variations in water supply, particularly during the bolting-flowering period. However, this reduction

remains closely linked to the nature of the genotype concerned. We can see that in the genotype it reaches a rate of 97%, whereas the average for all the genotypes is 9.5%.
The effect of the absence of irrigation on the development of pi length appears to be weak under these conditions. The average reduction rate reported for all genotypes was 4.48%. In the irrigated plot, ear length values ranged between extremes of 3.8cm±0.25 and 7.23cm±0.51, recorded respectively by the genotypes Sooty-9/Rascon-37/4/Storlom/3/Rascon37/... and CMH77.774/CORM//SOOTY_9/RASCON_37/3/SOMAT_4INTER_8. Whereas under rainy conditions these values are of the order of 3.3cm±0.1 and 6.7cm±0.44, recorded inMorder by genotypes SOMAT_4/INTER_8/4/GODRIN/GUTROS//DUKNEE_11/5/ and LANGLOIS.

Table 8- Effects of genotype, water regime and their interaction on the expression of morphological parameters measured during the 2014/2015 season.

	HP	LDE	LCE	LE	PMG
Genotypes	5.14***	5.17***	8.23***	2.44***	10,7608***
Water situation	287.56***	29.75***	33.70***	16.30 ns	1245,7230***
Genotypes*Water conditions	4.20***	3.40***	5.74***	1.69***	7,2953***

HP = Height of the plant, LDE = Length of the last internode, LCE = Length of the neck of the ear, LE = Length of the ear, ***= significant effect at p< 0.001.

1.2. Contribution of the various plant organs to grain filling over the three seasons

1.2.1. Campaign 2012/2013

During the 2012/2013 season, grain filling was essentially ensured by the current photosynthetic activity of the ear and the last leaf (Fig. 14). The contribution of these two organs was high in both treatments, with and without irrigation. The average contribution of the ear of all the genotypes tested was 23% in the irrigated lot and 19% in the rainfed treatment (Appendix Tab. 5). The contribution of the last leaf to grain filling was 16.13% in the irrigated lot and 17.23% in the rainfed lot. However, there was a very pronounced genotypic difference in the involvement of these two organs in grain filling. Regarding ear contribution ratios, in the irrigated treatment, the highest values of 44% and 40% were found in the ACSAD 1107 and Bousselem/Ofanto genotypes. The involvement of this organ in the dry treatment reached a maximum value of 43% in Ajaia-16//Hora/jro/3/Gan/4/Zar/5/Malmuk-l..... The relative contribution of the ear to the filling process reached values of 16.25% (irrigated conditions) where the Stj3/Bcr/Lks4/3/Ter-3 genotype stood out with a maximum value of 37.41%. Under

dry conditions, the average contribution is around 10.35% with a maximum value of   40,56% registered by the genotype NUS/SULA//5*NUS/4/SULA/RBCE_2/3HUI//CIT71/CII*2/5/.... Remobilisation of reserves is also an important source of grain filling. The average rates of this involvement are 11.21% (irrigated) and 15.36% (rainfed). These results indicate that in dry conditions, remobilisation of reserves is a more important source of the assimilates essential for grain filling than in irrigated conditions. In the irrigated plot, the average contribution of remobilisation reached a maximum value of 37.40% (ACSAD1339). In the rain-fed plot, the contribution of this source was 49.32% for genotype ACSAD1315.
During this season, the ear neck and the beard were also involved in grain filling, albeit to a lesser degree than the other sources. It should be noted that the role played by the ear neck in grain filling is more important in dry conditions, unlike that of the beard.
The average involvement of the cob neck was around 9.44% (irrigated) and 9.77% (rainfed). However, a distinction was noted among the genotypes for this contribution where it a reached 27%noted in the
MOHAWK/10/PLATA_10/6/MQUE/4/USDA573//QFN/AA_7/3/ALBA-D genotype under irrigated conditions, and 30% in the NUS/SULA//5*NUS/4/SULA/RBCE_2/3HUI genotype. //CIT71/CII*2/5/... in dry conditions.

The average relative contribution of the beard is 9.67% (irrigated conditions) where the Waha genotype stands out with the highest value evaluated at 26.42%. Under dry conditions, the contribution of the photosynthetic activity of the beard reached 9.31%, with the genotype BICHENA/AKAKI_7/3/SOMAT_3/PHAX_1//TILO_1LOTUS_4/7/... showing the highest rate at 29.17%.

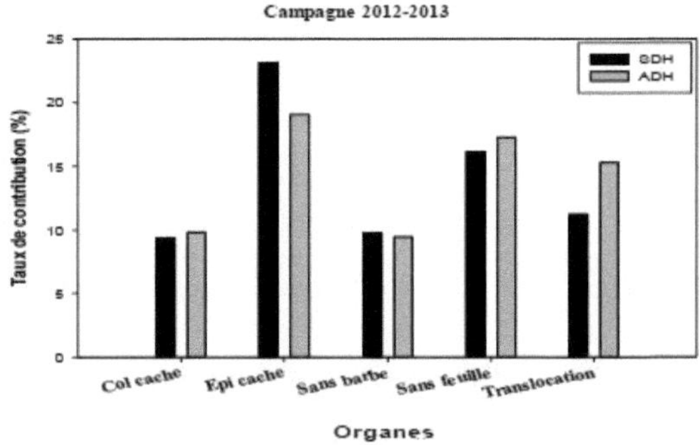

Figure 14- Relative contribution of excised and shaded organs to grain filling during the 2012-2013 season.

1.2.2.2013 campaign- 2014

During this season, the involvement of the different sources in grain filling is following different trends to those observed during the previous year (2012/2013). Under these conditions, there is a clear difference in the contribution rates of the different organs in this process with the variations in the water regimes carried out. Generally speaking, it is the remobilisation of reserves, the photosynthetic activity of the beard and the ear that are most involved in grain filling, particularly under dry conditions (Fig. 15).

Remobilisation of reserves makes the highest relative contribution to grain filling (Appendix Tab.6). The rates for this source were 11% (irrigated conditions) and 18.14% (rainfed conditions). The maximum values were around 41.81% (IA.1D5+10-6/3*Moja//Rcol/4/Arment//SRN_3/Nigris_4/3/Canelo_9.1) and 60.42% (GLOIRE DE MONGOLFIER) in the irrigated and dry treatments respectively.

The contribution of the photosynthetic activity of the ear was 12.57% (irrigated) and 11.39% (rainfed). The maximum contribution rates recorded in for the two water treatments are around 48.30% (OFANTO/WAHA) and 36.58% (CANELO_9.1/SNITAN/10/PLATA_10/6/MQUE/4/USDA573//QFN/...).

The photosynthetic activity of the beard makes a significant contribution to grain filling, particularly under dry conditions, where it reaches values of 8.75% (irrigated) and 14% (dry). Under rainfed conditions this contribution reached a maximum value of 41.61% (ACSAD1359), while under irrigated conditions this involvement reached 28.34% (LAHN/CHI2003).The average contribution of assimilates from the photosynthetic activity of the neck of the ear and the last leaf were 7.33% (irrigated), 9.89% (dry) and 7.86% (irrigated), 11.82% (dry) respectively. In the irrigated treatment, the maximum values for the involvement of the neck of the ear are around 26% (Acsad 1289) and in the dry batch this maximum is 33% (CANELO_9.1/SNITAN/10/PLATA_10/6/MQUE/4/USDA573//QFN/ ...). The maximum values shown by the relative contribution of photosynthetic activity in the latter are equivalent to 28.63% (Ajaia-16//Hora/jro/3/Gan/4/Zar/5/ Malmuk-1...) and 40.21% (CNDO/VEE//7*PLATA_8/3/GUANAY/10/PLATA_10/6/MQUE /4/...) recorded under irrigated and rainfed conditions respectively.

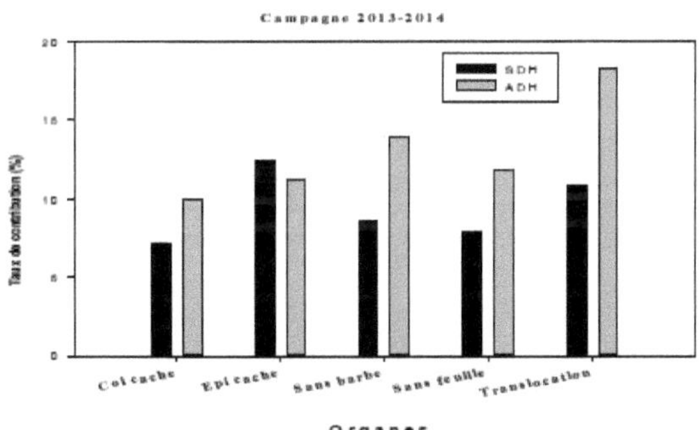

Figure 15- The relative contribution of excised and shaded organs in grain filling during the 2013-2014 season.

1.2.3. 2014 campaign- 2015

During this year's experiment, the contribution of the various organs involved in grain filling varied greatly among the genotypes and according to the water regimes applied. Generally speaking, grain filling relies equally on reserves from the five sources studied (Fig. 16).

The results obtained (Appendix Tab.7) show that the highest contribution rates come from the remobilisation of reserves, the photosynthetic activity of the ear and that of the neck of the ear. Remobilisation of reserves had average contribution rates of 10.48% (irrigated) and 17% (rainfed). Within this implication, the maximum values were of the order of 33.19% (OUED ZENATI) and 45.54% (Stot//Altar84/ALD) recorded respectively within the irrigated and rainfed treatments. The average contribution of the remobilisation of reserves increased by around 39% in dry conditions compared with irrigated conditions.The contribution of photosynthetic activity of the ear was 12.49% (irrigated) and 14.81% (dry). The maximum values for this contribution were 41.41% (CBC 514 CHILE/3/GUIL//GREEN) and 42.86% (Rascon_22/Rascon_21//Mojo_2/3/Guanay/4/Rcol/5/ Sora/2*plata_12/Somat_3) recorded under irrigated and rainfed conditions respectively.The photosynthetic activity of the neck of the ear also makes an appreciable relative contribution to grain filling, with average rates o f around 10.63% (irrigated) and 13% (dry). Under irrigated conditions, the highest contribution of this organ was noted in the ACSAD1367 genotype with 37%, while under rainfed conditions this optimum was achieved by the Stot//Altar84/ALD genotype with 48%. The contribution of the last leaf and the beard is higher in dry conditions during this year's experiment. The average involvement of the last leaf in the irrigated lot was 10.41%, with the ACSAD1339 genotype holding the highest value at 37%. Under rainfed conditions, the average contribution of this organ was 14%, with a maximum value of 43.39% recorded for the Ajaia- genotype.16//Hora/3/Gan/4/zar/5/Sooty-9....The results obtained from previous studies on the involvement of the beard in grain filling remain controversial. The data obtained in the present work indicate that the beard has an average share in grain filling of around 8.80% and 12.42% under irrigated and rainfed conditions respectively.The maximum values of this involvement observed in the irrigated and dry lots, equivalent to 33.4% and 41.55%, were recorded by Waha and the Llareta Inia/Guanay//Rascon-37/2*Tarro-2 genotype respectively.

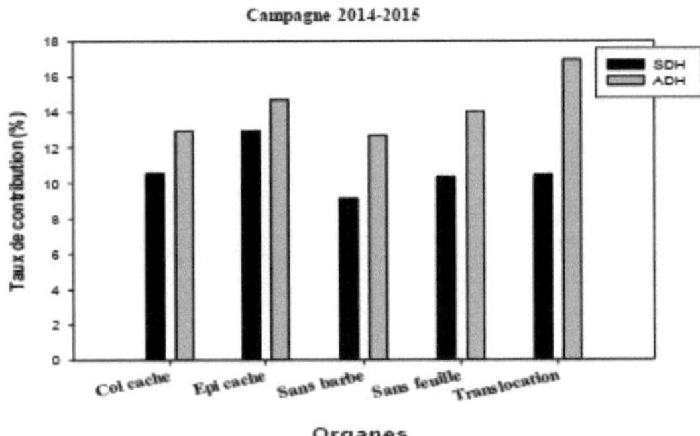

Figure 16- The relative contribution of excised and shaded organs in grain filling during the 2014-2015 season.

DISCUSSIONS

Grain filling and ripening is the last vegetative phase, and corresponds to the development of the final component of yield, which is the weight of the grain, following the migration of substances produced and stored by the plant's various components (Abbassenne et al., 1998; Gate, 2003). Grain filling and its quality depend on the expression of a set of parameters linked to the course of this process. Grain filling depends on two sources of assimilates, current photosynthesis and the remobilisation of reserves stored in the reserve tissues of the stem organs. Current photosynthesis, which is responsible for the availability of photoassimilates essential for grain filling, depends on the longevity and location of the organs involved.Various studies (Arous et al., 2020) indicate that in durum wheat, the last leaf, the ear beard, the ear husks and the ear neck play an important role in filling the grain and determining its final weight and quality. The availability of assimilates from the remobilisation of reserves depends on the morphology of the plant. However, the results obtained from the present study indicate that the contribution of the various sources to grain filling depends greatly on the quality of the plant's water supply and the nature of the genetic variability tested (genotypes). The effects of water supply are justified by changes in the size and longevity of the various morphological traits involved in this process. This study shows that water deficit reduces the magnitudes of all these parameters and reduces the longevity of certain organs that are sensitive to this environmental factor. This finding was observed between the two water treatments and between the years of experimentation, which were characterised by quantitative and spatial variability in the rainfall recorded.The remobilisation of stored reserves, particularly in the various parts of the stem, plays a major role in grain filling (Bahlouli et al., 2008; Belkharchouche et al., 2009). This contribution is higher in dry conditions, where photosynthetic activity is inhibited by poor water conditions. Morphologically, high-straw genotypes are well suited to this source of grain filling. These results are borne out by the work of Blum (1988), Monneveux (1992), Rebetzke et al (2002), Annichiarico et al (2005) and Ehdaie et al (2006), who have shown that durum wheat genotypes with high straw content and, consequently, late phenology produce grains with a higher weight, mainly under poor water supply conditions. This is mainly due to an increase in the following parameters reserve accumulation sites and the positioning of certain organs, such as the ear neck, adjacent to the developing grain. According to Bahlouli et al (2008) and Wardlaw (2002), grain filling, particularly in dry conditions, is essentially ensured by the remobilisation of reserves rather than by post-anthesis photosynthesis activity, which is affected by the scarcity of available water resources.The increase in plant height could also play an indirect role in grain weight, since this characteristic is associated with the presence of a deep root system, helping to avoid the stresses of drought.Through their role in photosynthesis and the production of assimilates, the components of the ear are necessary for grain filling. In the event of a water deficit, photosynthesis in the ear contributes relatively more to grain filling than the flag leaf (Johnson and Moss, 1976; Gate et al., 1993; Bammoun, 1997). This contribution comes from the photosynthetic activity of the grain envelopes and from the grain itself during its formation. As a result, an increase in the length of the ear contributes to an increase in its photosynthetic capacity (Kahali, 1995; Khan et al., 2010; Maydup et al., 2010; Maydup et al., 2014).The involvement of the last leaf in the development of grain quality in durum wheat has been proven by numerous studies (Monneveux et al., 2006; Evans et al., 1996). Through its photosynthetic activity during grain filling, the leaf ensures the availability of photoassimilates which, after transformation, accumulate in this organ. The results of this study show that the last leaf makes a significant contribution (Figs. 14, 15 and 16). This is particularly true in the case of batches grown under water deficit conditions. Although water deficit accelerates leaf senescence, this contribution is indirect, with the leaf's activity supporting the constitution of reserves during its activity and through the translocation of carbon and nitrogen reserves resulting from the reduction in its longevity (Amigue et al., 2006). The work of Tambussi et al (2005) and Calderini et al (2000) shows that the contribution of the last leaf is much higher than that of the leaves of the other stages of the plant.

2. Tests under controlled conditions

2.1. Trial 1: Contribution of organs to the filling of grain

2.1.1.Morphological characterisation of the genotypes used

The main morphological parameters of the genotypes used to assess the process of grain quality development were measured (Tab. 9). The results show that plant height, length of the last node, length of the neck of the ear, length of the ear, leaf area and length of the beard are significantly different among the genotypes studied ($p<0.001$). With regard to plant height, the genotypes tested can be divided into two groups. The first group includes the Langlois and Oued Zenati genotypes, which are said to have high straw height, with values of 148.07cm and 147.51cm respectively. In the second group are the short-straw genotypes with values of 94.68cm (Waha), 87.84cm (Mexicali75) and 85.65cm (ACSAD1361). As for the length of the last knot, the Langlois and Oued Zenati genotypes have the highest values with 44.28cm and 43.80cm respectively. The length of the neck of the ear followed a different trend, with the Waha and Mexicali75 genotypes having the longest necks, with data in the order of 22.67cm and 22.43cm.In certain environments, the last leaf plays a key role in grain development. The genotypes used in our experiment produced leaf areas of different values. The areas recorded vary between 46.57cm² (Langlois) and 34.94cm² (ACSAD1361). The other genotypes recorded intermediate surface areas with 39.71cm² (Oued Zenati), 38.39cm² (Waha) and 37.62cm² (Mexicali75).The length of the spike beard also varied between the genotypes studied. The longest beard was recorded in Langlois at 16.26cm. The genotypes ACSAD1361 and Waha had the shortest beards at 10.86cm and 10.58cm respectively. The other genotypes, Oued zenati and Mexicali75 showed intermediate lengths (Tab.9).

Table 9 - Average results for the morphological characteristics of the genotypes tested.

Genotypes	HP	LDE	LCE	SF	SF	LB
ACSAD 1361	85,65a	40,59a	21,65a	8,82b	34,94c	10,86a
Langlois	148,07c	44,28a	18,20b	9,98c	46,57d	16,26c
Mexicali75	87,84	40,25	22,43	8,35	37,62	12,66
Oued Zenati	147,51	43,80	22,32	8,33	39,71	13,27
Waha	94,68	40,33	22,67	8,47	38,39	10,58

HP: Plant height, **LDEN**: Internode length, **LCE**: Ear neck length, **LE**: Ear length, **LB**: Beard length, **SDF**: Last leaf area. Means followed by a different letter are significantly different according to Duncan's test at p<0.05.

2.1.2. Effect of water supply and sources of reserves on final weight and grain filling

The results show that the final grain weight depends on the nature of the genotype, the water supply and the source of accumulated reserves (Tab.10). In control plants (whole) and under optimum water conditions (100% CC), the variation between genotypes was about 11 mg, which represents almost a quarter of the average grain weight of all genotypes (Tab.11). In the same plants grown at 60% CC, grain weight fell significantly, with values ranging from 9% (Langlois) to 18% (Oued Zenati). This reduction was more marked at 30%FC, where the highest value was recorded by Oued Zenati at 39% compared with water conditions (100%FC). Under the same water conditions, the lowest value was recorded by ACSAD1361 at 21%.Excision or shading of the vegetative organs by the different treatments resulted in a significant reduction in final grain weight (Tab.11). However, this reduction depended on the genotypes tested and the levels of water supply. Among all the treatments, shading of the whole plant and shading of the ear resulted in the greatest reductions in grain weight for all three water treatments (Tab.11). This indicates that photosynthesis of the ears and remobilisation of reserves are the main sources of grain filling. Their respective average contributions at 100%CC were 38% and 52.36% respectively.These rates are more pronounced at 60% CC and 30% CC, reaching values of 41.8%, 54% and 46.1%, 53.7% respectively. The contribution of these two pathways to grain filling also differed between genotypes, particularly under conditions of 30% CC. Under this water regime (30% CC), the contribution of photosynthetic activity (Fig. 17) in the ear varies between 34.7% (Mexicali75) and 60.2% (Waha) and the contribution of remobilisation of reserves ranges between 42.7% (Langlois) and 64.1% (Mexicali75).

The contribution of the ear collar varies considerably depending on the genotypes and the water conditions adopted. Its contribution increases progressively under poor water conditions. Values of 9.2%, 14.5% and 19.8% were recorded in the 100% CC, 60% CC and 30% CC water treatments respectively. In the batch with 30% CC, the highest value of this contribution was recorded by Waha with 28.3%.

An opposite trend was noted for the contribution of the beard to grain filling, where it increased with the improvement in water supply conditions. The values recorded for this contribution are of the order of 13.6%, 12.9% and 10.9% respectively for the 100% CC, 60% CC and 30% CC treatments.

Lastly, the various leaf stages show contribution rates ranging from 14 to 27%. Among all the leaves, the flag leaf or the last leaf has a share of between 80 and 90%. The rates of its involvement in grain filling vary, according to the different water treatments (Fig.17), between 14.7% (100% CC) and 27.1% (30% CC).

Table 10 - Effect of genotypes, water supply and treatments on final grain weight and the relative contribution of the various organs to grain filling.

	Genotypes	Water situation	Treatments	Genotypes x Water situation	Genotypes x Treatments
Final grain weight	972,3***	2764,9***	2386,6***	32,2***	25,5***
Relative contribution	196,16***	4,03**	3209,02***	26,26***	46,87***

*** Significant at 0.001 probability; **Significant at 0.01; ^{ns} not significant.

Table 11- Average grain weight (mg) obtained in the different treatments and under the three water regimes (100%FC, 60%FC, 30%FC) for the five genotypes tested.

Genotypes	SH	CC	EC	BE	PC	DFE	EFE	PT
	30%FC	27,15	17,86	34,14	17,73	29,50	28,61	36,05
ACSAD1361	60%FC	31,59	19,78	35,98	19,30	31,14	31,01	39,29
	100%FC	37,06	29,31	36,63	25,29	35,22	34,65	45,48
	30%FC	25,80	14,31	29,80	17,54	29,99	27,21	36,01
Waha	60%FC	30,08	26,20	34,01	20.69	33,56	30,38	40,86
	100%FC	42,70	32,06	41,15	24,12	37,94	34,26	47,46
	30%FC	24,83	16,61	27,14	12,87	26,28	25,63	32,11
Oued Zenati	60%FC	41,28	23,94	35,18	15,20	32,68	33,61	43,27
	100%FC	51,12	30,39	39,40	15,35	36,57	36,22	52,73

	30%FC	36,77	26,26	36,75	23,92	39,66	36,72	41,75
Langlois	60%FC	45,48	34,36	46,32	29,18	41,75	40,62	51,11
	100%FC	52,72	38,02	54,10	31,42	45,03	42,93	56,43
	30%FC	29,55	21,85	29,35	11,98	27,46	26,99	33,43
Mexicali75	60%FC	38,05	23,08	37,84	16,08	34,77	33,00	42,84
	100%FC	44,12	25,23	44,95	22,37	37,84	36,72	47,82

(CC: Hidden collar; EC: Hidden ear; BE: Excised barbs; SP: Hidden plant; DFE: Last leaf excised; EFE: Excised leaf stage; PT: Control plant).

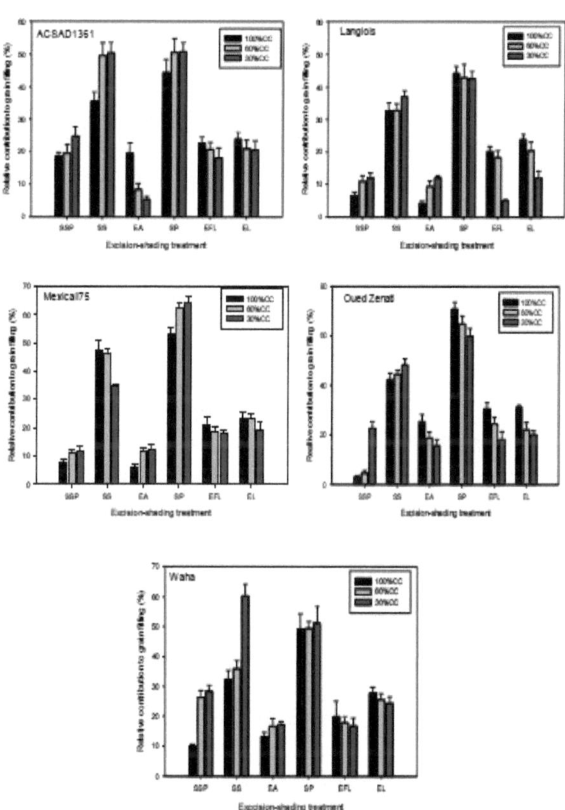

Figure 17- Relative contribution of the different organs to grain filling in five durum wheat genotypes under the three hydric levels adopted (30%CC, 60%CC, 100%CC).

DISCUSSIONS

The filling process determines the weight and quality of durum wheat grains. However, it is frequently subject to environmental constraints, particularly drought. The water deficit that occurs during the post-anthesis phases has negative effects on this process. For example, grain weight could be considerably reduced if plants are exposed to a short period of water deficit (Nezhad et al., 2012; Farooq et al., 2014). This trait (thousand kernel weight) has been reported as the most sensitive yield component to variations in water supply at the post-anthesis stage, in durum wheat (Ivanova and Tsenov, 2011). The results of these studies confirm those obtained in this work, which show that the increase in water deficit was accompanied by a significant reduction in the weight of a thousand grains at physiological maturity. However, this effect is linked to the nature of the genotypes grown. Application of the moderate water deficit (60% CC) caused a reduction in grain weight of 9%. Increased water deficit (30% CC) further reduced grain weight, with rates fluctuating between 39% (OuedZenati) and 21% (ACSAD1361). The effect of the water deficit on the final weight of the grain is mainly explained by a reduction in its size and a disruption in the migration of reserves. Thus, drought and high temperatures greatly reduce the size of durum wheat grain. This effect is explained by an inhibition of cell division and the organisation of the reserve accumulation tissue (Schnyder and Baum, 1992; Calderini et al., 2000). After differentiation of the seed reserve parenchyma, the effect of water deficit on its filling depends on the availability of assimilates and their migration towards the grain. This process therefore depends on the activity and longevity of the organs involved in the production and storage of photoassimilates. Remobilisation of stem reserves is the ultimate route to grain filling. Its relative contribution ranges from 44.05% to 70.49%. Variations in these values depend on the water supply and the genotypes tested. For Oued Zenati, for example, the relative contribution of the stem to grain filling increased as its water supply decreased (Fig. 17). For the other genotypes, the values were high, whatever the water regime applied. The role of stem remobilisation in grain filling has been reported in several studies (Cruz- Aguado et al., 2000; Tambussi et al., 2007), and Ehdaie et al. (2008) showed that the rate of this contribution varies between 10% and 50%, depending on the genotype and water supply. Other studies have noted that the stem contributes more to grain filling under drought conditions (Monneveux et al., 2006; Lopes et al., 2006).Shading of the ears led to a significant reduction in grain weight (Fig. 17). This is explained by its relatively important contribution to grain filling, which was second only to that provided by the remobilisation of stem reserves. This contribution is justified by its photosynthetic activity during grain formation and filling and by its proximity to the developing grain. This contribution is high under all three water conditions. However, in the ACSAD1361, Langlois and Waha genotypes, the relative contribution of the ear increased with decreasing water availability. In the Waha genotype, it was 32.37%, 35.91% and 60.21% in 100%, 60% and 30% CC respectively. The role of the ear is explained by its longevity (duration of photosynthesis activity), particularly under drought conditions. Indeed, Mohammady et al (2009) have shown that grain husks are less sensitive to the effects of drought, thanks to their greater osmotic adjustment capacity than other organs. Araus et al (1993) showed that in triticale, the different parts of the ear ensure the availability of a large proportion of photosynthate, which is essential for grain filling. The barbs are one

40

of the components of the ears whose relative contribution has been studied separately. Its involvement in grain filling remains relatively low and is highly dependent on genotypes and water supply (Abbad et al., 2004; Merah and Monneveux, 2015; Merah et al., 2018).

Our results showed that the contribution of the last leaf diminished as water supply levels fell (Fig. 17). In fact, the lowest contribution of the last leaf to grain filling was observed in the batch grown at 30%FC. This trend could be explained by the effect of water deficit, which reduces the longevity of this organ (Martinez et al., 2003; Merah et al., 2018).

The photosynthetic activity of the stem also ensures the availability of assimilates that are essential for grain filling. Among the various parts of the stem, the peduncle (spike neck) is of particular interest because its contribution is accentuated in dry conditions. Lingan et al (2010) suggested that the ear neck is a photosynthetically active organ that produces photosynthates and thus makes a decisive contribution to grain growth, particularly during the final stages of grain filling. In addition, Takahashi et al (2001) and Esmaeilpour-Jahromi et al (2012), reported that carbon assimilated during vegetative growth and early reproductive stages of wheat is temporarily stored in stem internodes and leaf sheaths, and can be remobilised and transported later to the developing grain.

Durum wheat (Triticum durum Desf.) is a cereal widely grown in semi-arid areas of Algeria, where it is subject to drought cycles that vary in intensity and timing. The occurrence of a post-anthesis drought is detrimental to grain filling and thus greatly reduces yield development. Grain filling, which leads to the development of its weight and quality according to its biochemical composition, depends on the availability of assimilates and their transformation into reserve compounds. The supply of substances comes from two sources: their remobilisation from stored reserves and the ongoing photosynthetic activity of the vegetative organs located above the last node.The results obtained from the present study are in line with others and show that there is a great deal of intra-specific variability in durum wheat, generating distinct behaviours for filling and quality development in situations of deficient water supply. These results provide numerous criteria that can be used in programmes to create cultivars that are tolerant to the constraints caused by late water deficit.Trials carried out over three consecutive seasons, under rain-fed and irrigated conditions, have shown that grain filling is achieved jointly by the remobilisation of reserves stored mainly in the stem and current photosynthesis in the vegetative organs located above the last node, particularly the neck of the ear, the ear, the beard and the last leaf. It has been shown that the share of each of these two sources in grain filling is dependent on water supply and the nature of the genotypes. Under optimum supply conditions, grain filling is essentially provided by the current photosynthetic activity of the last leaf and the constituents of the developing ear. Under these hydric conditions, remobilised reserves make a relatively smaller contribution than the first source. Under rainfed conditions, the opposite situation has been demonstrated, where grain filling is ensured mainly by assimilates from remobilised reserves and their translocation. The contribution of current photosynthesis particularly concerns the neck of the ear and the constituents of the ear in formation. This part of the work has shown that, in addition to the water factor, other climatic parameters, particularly temperature, have a significant effect on grain formation and quality. This is demonstrated by the significant inter-annual variations in the contribution of different sources to the availability of assimilates essential for grain filling. This indicates that plant morphogenesis and the differentiation of functional organs during this process are greatly influenced by climatic conditions. The effect of water deficit was distinguished from that of other climatic parameters by carrying out experiments under controlled conditions. At this level, by conducting the genotypes under three evolving water regimes (100%CC, 60%CC, 30%CC), it was possible to distinguish the action of drought on the relative contribution of the various plant organs to grain filling. Thus, the increased severity of the water deficit on the plant was accompanied by a clear contribution from the remobilisation of reserves, the photosynthetic activity of the ear and its collar in grain filling. Grain filling kinetics are also an important and easily detectable indicator of water deficit tolerance during grain formation. The evolution of dry matter deposition or the growth kinetics of the grain during formation depends on the longevity of the organ supplying assimilates. It should be noted that the duration of activity of the various organs varies and is conditioned by the plant's water supply. We also note from the present work that grain quality is conditioned by water supply, the nature of the genotype and the origin of the sources of assimilate influx at grain level. These factors are sources of major variations in the quality of starch deposited in the shape of the grain. The results obtained in the present work elucidate the relative contributions of the various plant organs to grain filling in durum wheat. They also provide information on the variations in these implications, imposed by climatic factors. However, further research is needed to assess the metabolic processes involved in transforming assimilates into reserve substances in the grain. The effects of the plant's water supply on the speed of protein synthesis and the polymerisation of simple sugars into starch need to be studied, as these processes are complementary to the synthesis of substrates for their actions, either stored or derived from the current photosynthesis of the plant's various organs.

REFERENCES

Abbassenne F., Bouzerzour, H. and Hachemi, L., 1998 - Phenology and production of durum wheat (TriticumdurumDesf.) in semi-arid zones. Ann. Agron. INA. N°18, p.24 - 36.

Abbad H., El Jaafari, S., Bort, J., Araus, J.L., 2004- Comparison of flag leaf and ear photosynthesis with biomass and grain yield of durum wheat under various water conditions and genotypes. Agronomie 24, 19-28.

Adda A, Soualem S., Labdelli A., Sahnoune M., Merah O., 2013. Effects of water deficit on the structure of the piliferous zone of durum wheat seminal roots. Revue écologie-environnement ,9. ISSN: 1112-5888.

Ali Dib T., &Monneveux, Ph., 1992- Adaptation à la sécheresse et notion d'idéotype chez le blé dur. I. Morphological rooting characteristics. Agronomie, 12 : 371-379.

Altenbach SB, DuPont FM, Kothari KM, Chan R, Johnson EL, Lieu D. 2003- Temperature, water and fertilizer influence the timing of key events during grain development in a US spring wheat. Journal of Cereal Science 37, 9-20.

Alvaro F., Royo C., Garcia del Moral L.F. & Villegas D., 2008 -Grain filling and dry matter translocation responses to source-sink modifications in historical series of durum wheat. Crop Science, 48: 1523-1531.

Amigues J.P., Debaeke P., Itier B., Lemaire G., Seguin B., Tardieu F., Thomas A. 2006. Réduire la vulnérabilité de l'agriculture à un risque accru de manque d'eau (Drought and agriculture). Expertise scientifique collective, synthèse du rapport, INRA (France), 72 p.

Annerose D. J. M. 1990-Research into the physiological mechanisms of adaptation to drought. Application to groundnut (Arachishypogea L.) grown in Senegal. Doctoral thesis in Natural Sciences, Université Paris VII, 282p.

AnnicchiaricoP., Bellah F., Chiari T. 2005-Defining sub regions and estimating benefits for a specific adaptation strategy by breeding programs: a case study. Crop Sci, 45: 1741- 1749.

Antoine C., Lullien-Pellerin V., Abecassis J. &Rouau X. 2002 - Nutritional interest of the wheat seed aleurone layer. Sciences Des Aliments, 22: 545-556

Araus J. L., Brown H. R., Febrero A., Bort, J. &Serret, M. D., 1993 -Ear photosynthesis, carbon isotope discrimination and the contribution of respiratory CO2 to differences in grain mass in durum wheat. Plant Cell & Environment, 16 (4), 383-392.

Araus J. L., Slafer G. A., Reynolds M. P. &Royo C., 2002 -Plant breeding and drought in C3 cereals: what should we breed for? Annals of Botany, 89 (7), 925-940.

Arous A., Adda A., Belkhodja M., Bouzid A and OthmaneMerah O., 2020 -The contribution of green plant parts to grain filling of durum wheat under water deficit Bulgarian Journal of Agricultural Science, 26 (N° 4), 809-815.

Auriau P., 1978- Sélection pour le rendement en fonction du climat chez le blé dur. Ann Argon d'El-Harrach. Vol8 N°2, 1- 14.

Ayed S., Carmous C ., Slim A., Amara H., 2010-Genetic variation of durum wheat landraces using morphological and protein markers. Africa journal of biotechnology, 49:8277-8282.

Bagga A.K., Ruwali K.N. and Asana R.D., 1970- Comparison of responses of some Indian and semi dwarf Mexican wheat to irrigated cultivation. Indian J, Agri, Sci, 40: 421- 427.

Bahlouli f., Bouzerzour H., Benmahammed A. AndHassous, K. L., 2008 -Selection of high yielding and risk efficient durum wheat (Triticum durumDesf.) cultivars under semi-arid conditions. Agro, 4: 360-365.

Bammoun A., 1997- Contribution à l'étude de quelques caractères morpho- physiologiques, biochimiques et moléculaires chez des variétés de blé dur (Triticumturgidumsspdurum.) pour l'étude de la tolérance à la sécheresse dans la région des hauts plateaux de l'Ouest Algérien. Magistère thesis, pp 1-33.

Barron C, Surget A, Rouau X. 2007- Relative amounts of tissues in mature wheat (Triticumaestivum L.) grain and their carbohydrate and phenolic acid composition. Journal of Cereal Science 45, 88-96.

Barron C., Abecassis S., Chaurand M., Lullien P., Mabille F., Rouau X., Sadoudi A., Samson M., 2012. Access to molecules of interest by dry fractionation. Innovation Agronomique, 19 :51-62.

Battais F., Rechard C., Leduc V., 2007- D'allergènes du grain de blé .revu française d'allergologie et d'immunologie clinique ,47 :171-174.

Battinger R., 2002- Arable crops. European Council of Young Farmers. 15p.

Bayuelo-Jiménez J.S., Graig R. And Lynch J.P., 2002- Salinity tolerance of Phaseolusspecies during germination and early seedling growth. Crop Sci. **42**, 1584- 1594.

Belitz, H.D., and Grosh, W. 1987. Cereal and cereal products. "Food Chemistry" Springer-Vergals. Berlin. Germany.

Belkharchouche H., Fellah S., Bouzerzour H., Benmahammed A., Chellal N., 2009- Vigueur de Croissance, Translocation Et Rendement En Grains du ble dur (TriticumDurumDesf) Sous Conditions Semi Arides. Courrier Du Savoir ,09 :17-24.

Benlarbi, M., Monneveux, P., Grignac, P., 1991- Etude des caracteres d'enracinement et de leur role dans l'adaptation au deficit hydrique chez le ble dur (TriticumDurumDesf.).Agronomie, 10; 305 - 313.

Benmahammed A., Bouzerzour H., Mekhlouf, A., &Benbelkacem A. 2008-Variation in relative water content, cell integrity, biomass and water use efficiency of durum wheat varieties (TriticumturgidumL. var durum) grown under water stress. Recherche Agronomique, INRA, 21: 37-47.

Benmahammed A., M. Kribaa, H. Bouzerzour, A. Djekoun. 2008- Relationships between F2, F3 and F4-derived lines for above ground biomass and harvest index of three barley (HordeumvulgareL.) crosses in a Mediterranean-type environment. Agricultural Journal, **3**: 313-318.

Biscope P.V., Gallagher J., Littleton E.J., Monteinth K.L. and Scott R.K., 1975- Barley and its environment. Sources of assimilates. J. Appel. Eco; 12: 395.

Blum A. 1988 -Plants breeding for stress environments. Boca Raton, 4, CRC, Press. Florida, USA. 223p.

Blum A., 1989 - Crop responses to drought and interpretation of adaptation. plant growth regulation 20, 135- 148

Blum A., 1996- Crop responses to drought and interpretation of adaptation. Plant Growth Regulation, 20:135-148.

Blum A., Pnuel Y., 1996- Physiological attributes associated with drought resistance of wheat

Blum A., Shpiler L., Golan G., Mayer J., Sinmena B., 1991-Mass selection of wheat for grain filling without transient photosynthesis. Euphytica54 : 111-116

Blum, A. 1998- Improving wheat grain yield under stress by stem reserve mobilization.

Euphytica, 100: 77-83.

Bogard M., 2011-Genetic and ecophysiological **analysis of** the deviation from the protein content - grain yield relationship in common wheat (TriticumaestivumL.). 37p.

Bootsma A., Boisvert J.B., Dejong R. &Baier W., 1996- Drought and Canadian agriculture. Sécheresse: 277 - 285 p.

Bort J., A Febrer O., Amaro T., Araus J.,1994- Role of awns in ear water-use efciency and grain weight in barley. Agronomie, EDP Sciences,14 :133-139.

Boufenar Z.F, Zaghoune O., 2006- Guide des principales variétés de céréales à pailles en Algerie (blé dur, blé tendre, orge et avoine) Ed. ITGC, 154P.

BoutignyL.,2007-Etude de l'effet de composés du grain de blé dur sur la régulation de la voie de biosynthèse des trichothécènes B : purification de composés inhibiteurs, analyse des mécanismes impliqués. 58 p.

Boyer J.S.,1982-Plant productivity and environment. Sci, New series. **218**: 443 - 448 p.

Branlard G., Pujos E., Nadaud I., Bancel E., Piquet A., 2012. New tools for fine analysis of grain composition. Innovations Agronomiques,**19**: 37-49.

Buleon, A., Colonna, P., Planchot, V., Ball, S., 1998-Starchgranules: structure and biosynthesis. International Journal of BiologicalMacromolecules 23, 85-112.

Calderini DF, Abeledo LG, Slafer GA, 2000- Physiological maturity in wheat based on kernel water and dry matter. Agronomy Journal 92, 895-901.

Carceller J.L, Aussenac T., 1999- Accumulation and changes in molecular size distribution of polymeric proteins in developing grains of hexaploidwheats: role of the desiccation phase. Australian Journal of Plant Physiology26, 301-310.

Chenafi H., A. Aïdaoui, H. Bouzerzour, A. Saci, 2006- Yield response of durum wheat (Triticum durum Desf.) cultivar Waha to deficit irrigation under semi-arid growth conditions. Asian J Plant Sci. **5**: 854-860.

Chinnusamy V., Zhu J. and Zhu J. K., 2006 -Gene regulation during cold acclimation in plants. PhysiologiaPlantarum 126 (1): 52-61.

CIC, 2013- Grain market, available at http://www.igc.int/, International Grains Council, accessed on 02/01/2014.

Clarck J. M.,Romagosa I.,Jana S.,Strivastava J .P . andMccaid T.N .,1989- Relation of excised leaf water lose rate and yield of durum wheat in diverse environments .Can, J. Plant .Sci ;69.P 1057-1081.

Clarke J.M., Townley - Smith, T.F., 1986 - Heritability and relationship to yield of excised leaf water retention in durum wheat. Crop. Sci. N° 26, p. 289 - 292.

Clarke, J.M., Norvell, W.A., Clarke, F.R. and Buckley, T.W., 2002- Concentration of cadmiumand other elements in the grain of near-isogenic durum lines. Can. J. Plant Sci./Canadian Journal of Plant Science, 82:27-33.

Clerget Y., 2011-Biodiversity of cereals, origin and evolution, 16 pages.

Colonna P, Buléon, A- 1992. New insights on starch structure and properties. In: Cereal

Chemistry and Technology: a Long Past and a Bright Future. Proceedings 9th International Cereal and Bread Congress, 25-42.

Condon A.G., Richards R.A., Rebetzke G.J., Farquhar G.D.. 2002 - Improving intrinsic water-use efficiency and crop yield. Crop Sci. 42: 122-131.

Condon, A. G., Richards, R. A., Rebetzke, G. J., Farquhar, G. D.,2004- Breeding for high water-use efficiency. Journal of Experimental Botany, Vol. 55, No. 407, Water-Saving. Agriculture Special Issue, PP. 2447- 2460.

Cooper P. J. M., Keating J. D. H., and Hughes G., 1983-Crop evapotranspiration-techniquefor calculation of its components by field measurements.Field Crops Res, 7: 299-312.

Cruz-Aguado J. A., Rodés, R., Pérez, I. P. & Dorado, M., 2000 - Morphological characteristics and yield components associated with accumulation and loss of dry mass in the internodes of wheat. Field CropsResearch, 66 (2), 129-139.

Debaeke P., Cabelguenne M., Casals ML. & Puech J., 1996-Élaboration du rendement du blé d'hiver en conditions de déficit hydrique. II. Development and testing of a simulation model for the cultivation of winter wheat under varied water and nitrogen supply conditions. Epicphase-blé. Agronomie.**16**: 25 - 46 p.

Debiton, C., Bancel, E., Chambon, C., Rhazi, L., Branlard, G., 2010- Effect of the three waxy null alleles on enzymes associated to wheat starch granules using proteomic approach. Accepted for publication in Journal of Cereal Science Ref: Ms. No. JCS10- 169R1.

Debiton C; 2011- Identification of wheat (Triticumaestivum L.) grain criteria favourable to bioethanol production by studying a set of cultivars and by proteomic analysis of isogenicwaxy lines.

Dupont F.M., Altenbach SB., 2003-Molecularand biochemical impacts of environmental factors on wheat grain development and protein synthesis. Journal of Cereal Science38, 133-146.

Ehdaie, B., Alloush, G.A., Madore, M.A., Waines, J.G., 2006- Genotypic variation for stem reserves and mobilization in wheat.IIPostanthesischangesin internode water- soluble carbohydrates. Crop Sci. 46, 2093-2103.

Ehdaie B., Alloush G. A. &Waines J. G., 2008- Genotypic variation in linear rate of grain growth and contribution of stem reserves to grain yield in wheat. Field Crops Research, 106 (1), 34-43.

Egli D.B., 1998- Seed biology and the yield of grain crops. CAB International, Wallingford, Oxfordshire, UK

El- Hakimi, 1995 - Physiological selection and use of tetraploid species of the genus Triticum for genetic improvement of drought tolerance in wheat.
Thesis of. Doctorat. Montpellier 220 pages.

Esmaeilpour-Jahromi M., Ahmadi, A., Lunn, J. E., Abbasi, A., Poustini, K. &Joudi, M., 2012 -Variation in grain weight among Iranian wheat cultivars: the importance of stem carbohydrate reserves in determining final grain weight under source limited conditions. Australian Journal of Crop Science, 6 (11), 1508-1515.

Evans G. W., Allen, K., Tafalla, R., & O'Meara, T.,1996- Multiple stressors. Journal of Environmental Psychology, 16, 147-174.

Evers T., Millar, S., 2002- Cereal grain structure and development: some implication for quality. Journal of Cereal Science 36, 261-284.

Farooq M., Hussain M. &Siddique K. H. M., 2014 -Drought stress in wheatduringflowering and grain-fillingperiods. CriticalReviews in Plant Sciences, 33 (4), 331-349

Feillet P., 2000- Le grain de blé composition et utilisation.1ère édition. INRA. Paris, 303p.

Feillet P., 2000- Le grain du blé : composition et utilisation. Mieux comprendre. INRA. ISSN: 1144-7605. ISBN : 2- 73806 0896-8.94-100.

Feillet, P., & Dexter, J. E. 1996- Quality requirements of durum wheat for semolina milling and pasta production, Monograph on Pasta and Noodle Technology, Saint Paul, MN; USA: AACC - American Association of Cereal Chemists, 95-131.

Fellah A., Benmahammed A., Djekoun A.et Bouzerzour H., 2002- Sélection pour améliorer la tolérance au stress abiotique chez le blé dur (TriticumdurumDesf.) .Actes de l'IAV Hassan II ,(Maroc) 22 , 161-170.

Ferero A., Rober J., Brown RH., Araus JL., 1990- The role of durum wheat ear as photosynthetic organ during grain filling. In: Advenced trends in photosynthetic, Malorca, Spain (unpublished).

Fischer R.A. and Maurer R., 1978 - Drought resistance in spring resistance wheat cultivar. I. Grain yieldresponses. Aust, J, Agri, Res, 29: 105-912.

Fokar M., Nguyen H .T., and Blum A., 1998- Heat tolerance in spring wheat II. Grain Filling Eupytica104 , 9 -15.

Foulkes M. J., Sylvester-Bradley R., Weightman R. &Snape, J. W., 2007 -dentifying physiological traits associated with improved drought resistance in winter wheat. Field CropsResearch, 103 (1), 11-24.

Gate P., Bouthier A., Casablanca H. and Deleens E., 1992 - Physiological characteristics describing the drought tolerance of wheat grown in France. Interpretation of correlations between yield and grain carbon isotopic composition. In: Tolérance à la sécheresse des céréales en zone méditerranéenne. Genetic diversity and varietal improvement. Montpellier (France) INRA. (Les colloques n°64).

Gate P., 1995- Ecophysiologie du blé, Edit. Lavoisier, Paris, Techniques et Documentations, 429.

Gate P., Bouthier A., Casabianca H., & Deleens E., 1993-Caractères physiologiques décrivant la tolérance à la sécheresse des blés cultivés en France : interprétation des corrélations entre le rendement et la composition isotopique du carbone des grains. Colloque Diversité génétique et amélioration variétale Montpellier (France). Les colloques. **64.**Inra. Paris.

Gharti-Chhetri, G.B. andLales, J.S., 1990-Biochemical and physiological responses of nine spring wheat (Triticumaestivum) cultivars to drought stress at reproductive stage in the tropic. Belg. J. Bot. 123, (1 / 2) : 27-35.

Godon B., et Wiliam C., 1991- Les industries des premières transformations des céréales. Ed Tec et Doc. Lavoisier.

Gooding M.J, Ellis R.H, Shewry P.R, Schofield J.D., 2003- Effects of restricted water availability and increased temperature on the grain filling, drying and quality of winter wheat.

Journal of Cereal Science 37, 295-309.

Hadjichristodoulou A., 1985- The stability of the number of tiller of barley varieties and its relation with consistency of performance under semi- arid conditions. Euphytica 34:641-649.

Hébrard J.P. 1996- Blé dur : objectif qualité, Nutrition : des pates épatantes. Document published for the symposium: perspectives blé dur, Toulouse, Labége, 26 November 1996 organised by : ITCF-ONIC-INRA-ITCF, p.6 - 7.

Hemery Y., Rouau X., Lullien-PellerinValérie, Barron C and Abecassis J., 2007- Dry processes to develop wheat fractions and products with enhanced nutritional quality. Journal of Cereal Science, vol. 46, pp 327-347

Hernandez J.A.Z., Santiveri F., Michelena A. And Pena R.J., 2004- Durum wheat (TriticumturgidumL.) carryng the 1BL/1RS chromosomal translocation: agronomic performance and quality characteristics under Mediterranean conditions. European Journal of Agronomy 30.

Hoseney R. C., 1994 - Principles of cereal science and technology. 2nd Edition, American Association of Cereal Chemists, 378 p.

Ishida .H, Yoshimoto.K, Izumi.M,Reisen D.,Yano.Y.,Makino.A., Ohsumi.Y.,Hanson M. R. and Mae T,2008- Mobilization of Rubisco and Stroma-Localized.

Ivanova N. &Tsenov N., 2011 -Winter Wheat productivity under favorable and drought environments' an overall effect. Bulgarian Journal of Agricultural Science, 17 (6), 777-782.

Jacquemin L., 2012-Production d'hémicelluloses de pailles et de sons de blé à une échelle pilote Etude des performances techniques et évaluation environnementale d'un agro- procédé. PhD thesis in Agroresources Sciences. Institut National Polytechnique de Toulouse (INP Toulouse). 345 pages.

Kamoshita, A., Chandra Babu, R., ManikandaBoopathi, N., Fukai, S., 2008- Phenotypic and genotypic analysis of drought-resistance traits for development of rice cultivars adapted to rainfed environments, Field Crops Research, 109: 1- 23.

Kehali L., 1997- Etude des paramètres d'élaboration du rendement hez le blé dur (TriticumdurumDesf.) Cultivées en conditions de déficit hydrique. Thèse de Magister. Université de Constantine.

Kellou R., 2008- Analysis of the Algerian durum wheat market and export opportunities for French cereals in the context of the Qualiméditeranée competitiveness cluster. The case of the Sud Céréales cooperatives. Thesis. Master. Centre international des hautes études agronomiques méditeraniéennes, Montpellier,159p.

Kent N, Evers A. 1994-Technology of Cereals. Technology of Cereals. 4thEdn, Pergamon Press, Oxford, UK. 109-119.

Khaliq, I., Irshad, A., Ahsan, M., 2008-Awns and flag leaf contribution towards grain yield in spring wheat (Triticumaestivum L.). CerealRes. Comm. 36, 65-76.

Khaldoune A., Kelkouli M., Kahlerras S., Amroune R. 1997. Techniques d'irrigation de complément des céréales en Algérie, brochure published by ITGC: 20p.

Langridge P.; Paltridge N. and Fincher G., 2006-Functional genomics of abiotic stress tolerance in cereals. Briefings in Functional genomics and Proteomics.(4): 343-354.

Lawlor, D.W., Day, W., Johnston, A.E., Legg, B.J. and Parkinson, K.J. 1981- Growth of spring barley under drought: crop development, photosynthesis, dry matter accumulation and nutrient content. Agric. Sci. Camb. 96; 167 - 186.

Li Y.H., Wang Q.J. & Ma A., 2003 -The osmotic adjustement and photosynthesis of wheat cultivar Hanfeng9703 with high yield, drought resistance under drought stressl, ActaAgronomica, 30: 759-764.

Lonbani M., and Arzani, A., 2011 - Morpho-physiological traits associated with terminal droughtstress tolerance in triticale and wheat. Agron. Res. 9, 315-329.

Lopes, M.S., Cortadellas, N., Kichey, T., Dubois, F., Habash, D.Z., Araus, J.L., 2006- Wheat Nitrogen metabolism during grain filling: comparative role of glumes and the flag leaf. Planta 225, 165-181.

Madhava Rao K.V., Raghavendra A.S. &Janardhan Reddy K., 2006 - Printed in the Netherlands. Physiology and Molecular Biology of Stress Tolerance in Plants. Springer: 1-14 p.

Marouf A. and Reynaud J., 2007-La botanique de A à Z. 1662 definitions. Ed Dunod : P.286.

Maydup M. L., Antonietta, M., Guiamet, J. J., Graciano, C., López, J. R. &Tambussi, E. A. 2010 - The contribution of ear photosynthesis to grain filling in bread wheat (Triticumaestivum L.). Field Crops Research, 119 (1), 48-58.

Maydup M. L., Antonietta, M., Graciano, C., Guiamet, J. J. and Tambussi, E. A. 2014- The contribution of the awns of bread wheat (Triticumaestivum L.) to grain filling: Responses to water deficit and the effects of awns on ear temperature and hydraulic conductance. Field Crops Research, 167: 102-111.

Mefti M., Abdelguerfi A. and Chebouti A., 2000-Etude de la tolérance à la sécheresse chez quelques populations de Medicagotruncatula(L.) Gaertn. Options Mediterraneennes HEAM, 173-176.

Mekhlouf A., Bouzerzour H., Benmahammed A., Hadj Sahraoui A., Harkati N. 2006- Adaptation of durum wheat (TriticumdurumDesf.) varieties to a semi-arid climate. Sécheresse. 17(4): 507-13.

Merah, O. &Monneveux, P., 2015- Contribution of different organs to grain filling in durum wheat under Mediterranean conditions. I-Contribution of post-anthesis photosynthesis and re-mobilization. Journal of Agronomy and Crop Science, 201 (5), 344-352.

Merah, O., Evon, P. &Monneveux, P., 2018 -Participation of green organs to grain filling in Triticumturgidum var. durum grown under Mediterranean conditions. International Journal of Molecular Sciences, 19 (56), 1-14.

Mogensen, V.O., 1991-Growth rate of grains of barley in relation to drought. Acta. Agric. Scand. 41: 345 - 353.

Mogensen, V.O., and Jensen, H.E., 1989- The concept of stress days in modelling crop yield response to water stress. Proceeding of the CE.C Worksop: Management of water resources in cash crops and in alternative production systems. Brussels 1988, 13 - 22.

Monneveux P., Nemmar M., 1986- Contribution to the study of drought resistance in common wheat. Study of proline accumulation during the development cycle. Agronomie 06 :583-590.

Monneveux P., 1991- Quelles stratégies pour l'amélioration génétique de la tolérance au déficit hydriques des céréales d'hiver. I n : l'amélioration des plantes pour l'adaptation aux milieux arides. AUPELF-UREF. Ed. John Libbey. Eurotest .pp: 165- 186.

Monneveux P., Saint-Pierre C. A.. et Comeau A., 1992-BarleyYellowdwarf virus tolerance in droughtsituations .ByDv in west Asia anâ North Africa . Proceeding of a workshop
.Rabat, Morocco,lg-21 November 1989.P:209-22

Monneveux, P., This, D. 1997- La génétique aux problèmes de la tolérance des plantes cultivées à la sécheresse : espoirs et difficultés. Sécheresse. 1(8) : 29-37.

Monneveux P., Rekika, D., Acevedo, E. and Merah, O., 2006 - Effect of drought on leaf gas exchange, carbon isotope discrimination, transpiration efficiency and productivity in field grown durum wheat genotypes. Plant Science, 170(4): 867-872.

Morgan, M. and Condon, A.G., 1986- Water use, grain yield and osmoregulation in wheat.

Aust. Plant Physiol. 13, 523 - 532.

Morrison WR,Gadan H., 1987- The amylose and lipid contents of starch granules in developing wheat endosperm. Journal of Cereal Science 5, 263-275.

Mosiniak M., Prat P., and Roland J.C., 2006-Biology and multimedia.Université Pierre et Marie Curie.

Naville M., 2005- La biodiversité des espèces cultivées : Analyse dans le cas du blé, Paris: Université Paris XI, Paris, 20p.

Nemmar M., 1980-Contribution à l'étude de la résistance à la sécheresse chez le blé tendre (TriticumastivumL .) : Étude de l'accumulation de la proline sous l'effet du stress hydrique .Thèse D.A.A. ENSA. Montpellier. 65p.

Nezhad K. Z., Weber, W. E., Röder, M., Sharma, S. S., Lohwasser, U., Meye, R. C., Saal,

B. &Börner, A., 2012 - QTL analysis for thousand-grain weight under terminal drought stress in bread wheat (Triticumaestivum L.). Euphytica, 186 (1), 127-138.

Olsen OA., 2001-Endosperm development: cellularization and cell fate specification. Annual

Review of Plant Physiology and Plant Molecular Biology52, 233-267.

Panozzo J.F., Eagles H.A., Cawood R.J., Wootton M.,1999- Wheat spike temperatures in relation to varying environmental conditions. Australian Journal of Agricultural Research 50, 997-1005.

Passioura J., 2004- Increasing crop productivity when water is scarce: From breeding to field management In: Proceeding of the 4thInternational Crop science Congress. New direction for a diverse planet. Brisbane, Australia. PP: 12, www.regional.orgau/au/cs.

Pereyra M., Del M., Torroba C., 2003-Molecular and physiological responses to water deficit in wheat (Triticum aestivum L.). Agronomia - UNLPamVol.14; 2p.

Rebetzke G. J., Condon R.A., Richards R.A., & Farquhar G.D., 2002-Selection for reduced carbon isotope discrimination increases aerial biomass and grain yield of rain- fed bread wheat, Crop Sci., 42: 739 -745.

Richards, R.A., 1983- Manipulation of area and its affects on grain yield in drought wheat.

Aust.]. Agric. Res. 34: 23 - 31.

Richards, R.A., Rebetzke, G.J., Van Herwaarden, V.F., Dugganb, B.L. and Condon, A.G. 1997-Improving yield in rainfed environments through physiological plant breeding. Dryland Agriculture N° 36, p. 254 - 266

Saeedipour S., &Moradi, F. 2011 -Effect of drought at the post-anthesis stage on remobilization of carbon reserves and some physiological changes in the flag leaf of two wheat cultivars differing in drought resistance. Journal of Agricultural Science, 3, 81-92.

Sarda, X., Vansuyt, G., Tousch, D. ,Cassedelbart, F. , et Lamaze, T., 1992- Les signaux racinaire de la régulation stomatique. In " tolerence a la secheresse des cereales en zones mediteranneenes '. Ed. INRA. Paris. 1993. 75 - 79.

Saulinier Luc, 2012- Les grains de céréales diversité et composition nutritionnelles. Cahiers de nutrition et diététique ,47 : 4-15.

Schnyder H. & Baum, U., 1992 - Growth of the grain of wheat (Triticumaestivum L.): the relationship between water content and dry matter accumulation. European Journal of Agronomy, 1 (2), 51-57

Sestili F., Janni M., Doherty A., Botticella A., D'ovidio R., Masci S., Jones H.D., Lafiandra D., 2010-Increasing the amylose content of durum wheat through silencing of the SBlla genes. Plant Biology 10 :144.

Shewry P.R, Tatham A.S, Forde J, Kreis M., Miflin B.J., 1986- The classification and nomenclature of wheat gluten proteins: A reassessment. Journal of Cereal Science 4, 97-106.

Shewry P.R., Underwood C., Wan Y., Lovegrove A., Bhandari D, Toole G., Mills E.N.C, Denyer K., Mitchell R.A.C., 2009- Storage product synthesis and accumulation in developing grains of wheat. Journal of Cereal Science 50, 106-112.

Simane B., Peacock J.M. and Strick P.C., 1993 - Differences in development plasticity growth rate among drought .resistant and susceptible cultivars of durum wheat (Triticum turgidumL.Var . durum) . Plant and soil, 157, 155- 166.

Singh J. and Skerritt, J. H. 2001 - Chromosomal control ofalbumins and globulins in wheat grain assessed using differentfractionation procedures. J. Cereal Sci.33: 163181.

Singh B.K., Jenner C.F., 1982 - Association between concentrations of organic nutrients in the grain, endosperm cell number and grain dry weight within the ear of wheat. Australian Journal of Plant Physiology 9, 83-95.

Slama A., Ben salem M., Ben Naceur M., Zid E., 2005- Les céréales en Tunisie : production, effet de la sécheresse et mécanismes de résistance. Article De Recherche, 16: 225-9.

Sofield I., Wardlaw I.F., Evans LT., Zee SY., 1977- Nitrogen, phosphorus and water contents during grain development and maturation in wheat. Australian Journal of Plant Physiology4, 799-810.

Soltner, 1988 - Les grandes productions végétales. Les collections sciences et techniques agricoles, Ed. 16éme éditions 464P.

Sorrells M.E., Diab A., Nachit M., 2000- Comparative genetics of drought tolerance. Mediterranean Options Series A (Mediterranean Seminars), 40: 19 1-201.

Sramkova Z., Gregova E.,Ernest S., 2009-Chemical composition and nutritionalquality of wheat grain. ActaChimicaSlovaca,volume(2) n°1: 115-138.

Stamova B.S., Laudencia-Chingcuanco D, Beckles D.M.,2009-Transcriptomic analysis of starch biosynthesis in the developing grain of hexaploid wheat. International Journal of Plant Genomics 2009, 1-23.

Stevenson D. G., Jane, J. L., and Inglett, G. E. 2007- Structure and physicochemical properties of starches from sieve fractions of oat flour compared with whole and pinmilled flour." Cereal Chemistry, 84(6), 533-539.

Stoddard F.L., 1999- Variation in grain mass, grain nitrogen, and starch B-granule content within wheat heads. CerealChemistry 76, 139-144.

Surget A,,Barron C., 2005- Histologie du grain de blé, Industrie des céréales 145, 4-7.

Taiz L. and Zeiger E., 2002- PlantPhysiology. 3rd ed. Sinauer Associates Publishers,Sunderland, 427 p.

Takahashi T., Chevalier P. M. & Rupp, R. A., 2001 - Storage and remobilization of soluble carbohydrates after heading in different plant parts of a winter wheat cultivar. Plant Production Science, 4 (3), 160-165.

Tambussi E A, Nogués S, Araus J L., 2005 - Ear of durum wheat under water stress: Water relations and photosynthetic metabolism. Planta, 221, 446-458.

Tambussi E.A., Bort J., Guiamet J.J., Nogués S., Araus J.L., 2007-The photosynthetic role of ears in C3 cereals: metabolism, water use efficiency and contribution to grain yield. Crit. Rev. Plant Sci. 26, 1-16.

Tardieu F., 2005- Plant tolerance to water deficit: Physical limits and possibilities for progress. Geo SCi, 337: 57 -67.

Tardieu F., 2003- Virtual plants: modeling as a tool for the genomic of tolerance to water deficit, Trends in Plant Sciences, 8: 9 -14.

Triboï E., 1990-Modèle d'élaboration du poids du grain chez le blé tendre.Agronomie. **10** : 191- 200p.

Tsimilli-Michael M. M., Pecheux R.J.et Strasser, 1998- Vitality and stress adaptation of the symbionts of coral reef and temperate foraminifers probed in hospiteby the fluorescence kinetics O-J-I-P. Archs. Sci. Genève.51: 205 - 240 p.

Vensel, W.H., Tanaka, C.K., Cai, N., Wong, J.H., Buchanan, B.B., Hurkman, W.J., 2005- Developmental changes in the metabolic protein profiles of wheat endosperm. Proteomics 5, 1594-1611.

Verville J.L., 2003. Wheat, rye and triticale. 18p.

Wardlaw I.F., 1967- The effect of water stress on translocation in relation to photosynthesis and growth. I. effect during grain development in wheat. Aust. J. Biol. Sci ; 20:25-39.

Wardlaw J.F., 2002- Interaction between drought and chronic high temperature during kernel filling in wheat in a controlled environment. Annals of botany, 90: 469-476.

Wiegand, C. L., Cuellar, J.A. 1981- Duration of grain filling and kernel weight of wheat as affected by temperature. Crop Sci. 21(1): 95-101.

Wilson J.D., Bechtel D.B., Todd T.C, Seib PA, 2006- Measurement of wheat starch granule size distribution using image analysis and laser diffraction technology. Cereal Chemistry 83, 259-268.

Witcombe J. R, Hollington P. A., Howarth C. J., Reader, S., and Steele, K. A., 2008 - Breeding for abiotic stresses for sustainable agriculture. Phil. Trans. R. Soc. B 363 : 703-716.

Yang J. and Zhang, J. 2006- Grain filling of cereals under soil drying. New Phytologist, 169(2): 223-236.

Yin X, Guo W, Spiertz JH. 2009- A quantitative approach to characterize sink-source relationships during grain filling in contrasting wheat genotypes. Field CropsResearch114, 119-126.

Zaddem, M., 2014- Application of the response surface method for the optimization of hydrogen peroxide bleaching of wheat bran and its incorporation into bread flour. Thesis. PhD. Université laval Canada. 110 p.

I want morebooks!

Buy your books fast and straightforward online - at one of world's fastest growing online book stores! Environmentally sound due to Print-on-Demand technologies.

Buy your books online at
www.morebooks.shop

Kaufen Sie Ihre Bücher schnell und unkompliziert online – auf einer der am schnellsten wachsenden Buchhandelsplattformen weltweit! Dank Print-On-Demand umwelt- und ressourcenschonend produziert.

Bücher schneller online kaufen
www.morebooks.shop

Printed by Books on Demand GmbH, Norderstedt / Germany